THE MANDRAKE EFFECT

Other titles by author

THE MANDRAKE EFFECT
THE AMERICAS
SEVEN VEILS OF WISDOM
THE COLEBRIDGE COVENANT

The British Collection
AMBASSADOR TO EARTH
THE FAIRFAX LEGACY
DUCKS N' DRAKES
KEEPING SCHTUM
AN INTERVIEW WITH…

pjsearlebooks@gmail.com

P J SEARLE

THE MANDRAKE EFFECT

© P J Searle
All rights reserved
2019

This is a work of fiction. Names, characters, places,
and incidents are the product of the author's imagination,
or based on rigorous historic research.

For Lew and Stew...

... and any of you that ever wondered what proverbial black hole the American manned-space-program disappeared into after the last moon landing.

BLOC ONE – SYNDROME

Personally I don't like introductions, synopses, or forewords. When I do knuckle-down to read a book – which isn't as often as I'd like – I like to get straight to the chase. But having said this I feel compelled to make two salient points:
—Point one: the Rolls-Royce turbo-jet debacle; the folly of the British government gifting to Russia, back in 1946, the then revolutionary Rolls-Royce Derwent/Neme turbo-jet technology. As a direct result the MiG-15 jet fighter, on its début in the Korean War, was to shock the West and chase all piston-engine fighter aircraft from the sky—
 If you are accustomed to a more, shall we say, 'spicy read' please bear with... I assure you it's not all unromantic geeky stuff.
—Point two: Project Orion; a feasibility study in 1958 post-war America to the development of NPP, *Nuclear Pulse Propulsion*. It was the first such think-tank of its kind since World War Two's Manhattan Project, which, as we all know, gave us the atom bomb. Cheap interplanetary space-travel was Orion's ultimate goal. But the Test Ban Treaty of 1963 is generally acknowledged to have ended the project—
 So, precious precious reader, male, female, geek or otherwise, please read on.

Put out the light, and then... put out the light:

Othello

CHAPTER ONE

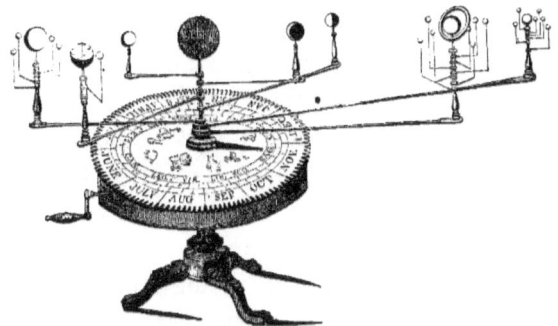

1952 – SPACE, CLOSING MARS ORBIT. A distant man-made eruption of gas appears rudely interrupting God's infinite serenity. At its epicenter a solitary projectile, a crude, unsophisticated cylindrical craft blowing out an enormous shroud of deadly radioactive propulsion material as it closes on the red planet, leaving behind the claustrophobic ever-present aura of timelessness.

Suddenly, inexplicably, a beam of white incoherent light floods the craft. Something from outside, something alien has pierced the integrity of the ship. This celestial light, after an incalculable passing of time, shifts through the spectrum to deep indigo.

The spacecraft continues its journey and successfully rounds Mars. Once clear of the planet's orbit the massive engines ram the projectile onwards at incredible speed away from the Red planet and on into space, and into the last faze of its elliptical parabola taking it towards its origin, to the only colour in the otherwise blackness, the planet Earth.

******

1980 – SPACE, CLOSING EARTH ORBIT. An age has passed. Now, into the unchanging indigo emptiness, a small gimlet-point of light appears. After a few moments the luminous speck has grown to a magnificent spacecraft, the United States Starship *Junairo*. This colossal craft glides slothfully onward to make rendezvous with the only other man-made body in this infinity, EarthlabOne, a huge elliptical space station. This orbiting oasis carries supplies, rocket fuel, and a permanent crew of twenty souls plus four SBS, *stand-by shuttle,* tethered beside like fledgling offspring huddled to their mother.

After various maneuvers the spacecraft gracefully docks and makes fast to the station. Her maiden test flight successful, Junairo will now undergo the laborious month-long procedure of refit.

Refit completed: Junairo is fuelled and crewed and towed one hundred nautical miles clear of the space station by a tug SBS. Duty done the shuttle detaches and returns to the station. After the mandatory countdown the starship fires its massive engines. An incredible eruption of energy rams the elegant craft into deep space and to its maximum velocity, one-third the speed of light, and on towards its ultimate destination, the planet Mars.

The interior of the spacecraft is lit to half-light, dimly illuminating the three glass topped hibernation coffers, occupied by two men and one woman. Next to these are three pressure suits, labeled, *Fizz, Lenny,* and *Rose.* The rest of the cabin is plain and uncluttered, totally devoid of any visible computer peripheral, apart from wall-mounted monitor screens. All is serene, inert and silent, just the overwhelming presence of timelessness. A week, a month, a year could pass, it would make little difference to the occupants of the coffers as they breathe once every five minutes, the breath of the near dead. In fact two months have flickered away unnoticed by the comatosed crew. Contrariwise, to the ship's onboard computer, time is essence; it is the lifeblood of the Junairo.

Now, on cue, the planet Mars offers itself for view on the ship's main monitor screen, but no conscious eye observers its red lifeless image. Also on cue the flight-deck clatters awake, together with other flickering monitor and computer readouts—the USS Junairo has risen from its slumber.

Suddenly, inexplicably, a white incoherent light floods the craft. Something from outside, something alien has pierced the integral tranquillity of the ship.

******

EDINBURGH UNIVERSITY, SCOTLAND. Inside this ancient redbrick house-of-learning Henry Mandrake paces the well-worn corridors. He is of slender build, good-looking and in his mid-thirties. Even before he opens his thin-lipped mouth it is obvious he will possess a cut-glass English accent. He marches on with flamboyant *Fred Astaire* gait, his tutor's gown draped foppishly over his Savile-Row jacket, the lapel buttonhole of which adorning his customary sprig of mandrake, two violet blossoms surrounded by a cluster of blue-green leaves. Onwards he dramatically charges, oozing self-importance, past a pretty woman who has stuck her head out of one of the many doors, in vain attempt to engage him in covert conversation.

'Harry,' she hisses. 'Tonight, he's–' Harry ignores, and

carries on down the corridor to the small reception at the far end where, with much pomp and circumstance, he stops. Behind the desk is a seated woman secretary, by her side, impatiently waiting, is a huge villainous-looking man.

'Radcliff, I presume?' challenges Harry to the thug in impeccable Eton accent.

The man turns, pushes his ugly, pugilistic scarred, face into Harry's space and growls in guttural Gorbals brogue, 'Yoo, Mandrrrakk?'

'Yes, very nearly—Follow!' Harry marches off without waiting for further comment, just a beckoning finger over his shoulder. Radcliff follows close behind, into one of the offices. As the door starts to close, Harry hurls his words. 'Now then, what the devil do you mean by coming here? And for God's sake shut the bloody door... we don't want the world and its mother listening.'

The door slams under the belligerence of Radcliff's muscular shoulder. Through the obscured reed-glass panel the receptionist unwittingly observes the jumbled figure of Radcliff grabbing Harry, thrusting his fist into his chest and proceeding to pummel him against the wall whilst hurling a bile of unintelligible threats, curses, and blasphemies. This goes on for half a minute or so, then Harry is shoved away. The door crashes back; the receptionist startles. Radcliff emerges and menaces off towards the car park shrugging, twitching, and neck jerking, construing 'job done.'

After a few moments Harry gingerly appears. He straightens his displaced tie and ruffled lock of hair, offers the receptionist a nervous smile and then storms off back down the corridor. Halfway along the woman tries again to speak with him:

'Harry,' she hisses, cocking a well-plucked eyebrow, 'He's away all weekend.' As Harry continues on she adds a wanton, 'know what I mean?' leer. Without breaking pace he forges an accepting smile to the eagerly offered tryst and marches on, gown flowing.

Inside the spacecraft Junairo, a tall slimly built woman – Navigation officer, Rose Hawkins – stands with her back to the cabin wall. She is dressed in a light-blue cooling suit that has the nametag 'Rose' over her heart. She appears terrified and in a state of near panic. Held tightly in her hands is a laser cutting-lance, aimed at a figure just out of view. On the verge of hysteria she calls out to the craft's open microphone. 'Oh my God, I must be dreaming. Can you see this, Major?' Unintelligible static answers. She screams back at the mike. 'EarthLab, come in damn you. Jesus Christ, I must have the bends.'

The figure across the cabin moves and starts to talk – a woman's voice. 'Please listen... I'm in pain. I'm sick... I'm burning. My whole body is on fire. I need help. Please, I'm–'

'Move again, freak,' yells Rose, cutting her off mid-sentence, 'and I'll burn you myself.'

'No, don't. You've got that thing set to infinity... you'll hole the ship. We'll both die.'

'I don't care. I'll cut you... I'll blast you.'

The second woman backs off and puts her hands up to her face in anguish. 'Oh God, what's happening? I'm in pain. Please don't point that thing at me. Where are the men?'

'You fucking tell me. – What am I doing, swearing at a freak?'

'I'm not a freak. Where's Lenny?'

'You should know, you're wearing his suit. Tell me, freak. Or so help me I'll spread us both over a mile of space.'

The second woman calms slightly and puts her hands down to her sides. 'I don't know,' she answers submissively, 'and I am not a freak, I'm as scared as you. I'm sick. I'm hurting.' She turns and calls into the open mike. 'Major, are you getting this? Base, for Christ sake are you getting this? Answer damn you,

They stand motionless, glaring at each other. Eventually the static clears and a voice gradually constructs:

'... can you hear me? Answer Junairo, do, you, read, us? This is EarthlabOne Private Enterprise. Do you read us? – Over.'

'Where the hell have you been?' Rose screams into the open mike. There is silence for a few moments. She closes her eyes and waits.

'Calm down, Rose, calm down.' The crackly voice jolts her to attention. 'We have you... sound only, just a one-track audio link so we can't both talk at once, we're working on it, but we have you. – Over.'

'Where the hell have you been? Thank God. Get video. We have a problem. – Over.'

'Whatever the problem, Rose, we can work through it. Calm down, plug your unit in so we can monitor your–'

'Don't mess with me, Base...' interrupts Rose, but the two voices collide over the one-way link. After a moment it clears and she continues. 'Let me speak to Major, and get video. And get it quick, I don't think I can explain this with words. – Over.'

Major, a powerful military man in his sixties, sits

looking into a blank monitor screen on the main deck of EarthlabOne. He speaks with controlled composure into a hand-held mike. 'Rose, this is Major speaking. Now, for heaven's sake calm down and plug-in your unit, we need to monitor you, just normal procedure. Let me speak to Lenny. I want to–' He stops mid-sentence aghast. The static has cleared and the grainy black-and-white picture that now constructs on his monitor screen causes him to gasp in utter shock. 'What in hell. What in hell?' He turns and yells orders to the entourage technicians. 'Hold that picture. Cut the loop. Absolutely no coverage.' He yells back into the mike, 'Rose, for Gods sake what's happened?'

Major stares unbelievingly into his monitor – There is Rose still in her cooler suit, cutting-lance in hand, pointed at… a second *Rose*, dressed in a similar cooling suit that has 'Lenny' over the heart. Even with the low-res picture it is unmistakably two identical women, both rigid with terror, both staring unbelievably at each other. At the back of the cabin, between the two women, are the three life-support coffers, one intact, the other two shattered. Around them is a pile of decomposing remains, the air littered with the lightest of this material floated up in the quarter Earth gravity of the ship. On the monitor the picture clears and sharpens into colour. After a long silence Major's voice, now loud and clear, rings around Junairo's cabin over the open radio-link:

'Here's what you do, Rose, suit up and–'

'I'm not suiting up. I'm not moving till you come up and get me.'

After another long silence, Major's voice shrills out again. 'For God sake, Rose, we can't–'

'Yes you can. Just do it. I'm not putting this lance

down, so you damn well get me out of here or I'll blast the freak.'

At EarthlabOne, Major runs his fingers through his balding hair. He stares unblinking into the monitor. 'We can't bring you in if you don't co-operate. Be reasonable, Rose.'

On Junairo, Rose re-aims the lance, her fingers still tight on the trigger. 'Reasonable,' she yells. 'Well here's reasonable; if you don't tell me you're sending an SBS I'll blast the lot to bits. My skin is crawling, my legs are shaking, and I've filled my urine-sack I'm that goddam scared. So don't give me fucking 'reasonable.''

'Rose, the shuttle blasted off fifteen minutes ago, the moment we picked you up on the scans. Now, for Christ sake will you calm down? Both of you, suit up, stabilise pressure, and close off the cabin.'

'No way,' yells Rose. She aims the lance and continues with cold determination, 'I'm not moving. She can suit up, but I'm holding this lance until you come through the hatch.'

'That won't do, Rose. Safety procedure, suit up and close off. We've got to flood your cabin. We're changing atmospheres for Christ's sake, air for gas... cadet training.'

'No fucking way.'

'If you don't close off there'll be a great rush of air and—'

'Don't mess with me.'

'Listen to me, Rose. You got to—'

'I mean it, Major. The men are rotted to slime and I'm left here with this freak goddam bitch nightmare staring at me. I don't know if I'm dreaming, hallucinating, or what, so just don't.' She turns and screams to the other woman,

'Stop fucking staring at me.' The woman diverts her eyes.

'Rose, let the other... person close off.'

'If she moves I'll burn her. You'd better start believing me, Major. I'll waste the freak and the whole damn ship.' She turns to the onboard camera, adjusts and points the lance upwards. A shaft of amber light streaks across the cabin and into an alloy purlin, sending a shower of molten metal droplets dancing gracefully across the cabin deck.

On EarthlabOne, Major views the action on his monitor. 'Oh, shit. ... Okay okay, give me a minute.' His voice muffles as he confers with his subordinates, then he comes through loud and clear, 'Okay Rose, here's what you do. The moment you hear them bang on the bulkhead, slip your helmet on with one hand and open the valves wide. It'll be hard to breathe – we don't have your cabin pressure – but you'll manage, breathe in gulps. You'll have to hold on to it real hard. When the airlock opens, the rush of air will probably knock you over, so don't have your finger on the trigger. Got that Rose? Rose. ROSE.. For Christ's sake, confirm.'

'Fucking got it. – *Freaking* got it, okay?' She closes her eyes in momentary prayer, 'Oh sweet Jesus, please help me. Please forgive me for swearing. Mary mother of God, help me, I'm so scared.'

They stare across the cabin, two beautiful women absolutely identical, absolutely terrified, drawn by the irresistible attraction of each other's eyes. Major's rasping voice breaks the leaden silence.

'They're coming about Rose. Helmet on.' Rose takes the glass dome helmet with one hand and, trailing the umbilical across the cabin deck, awkwardly slings it on.

'It's done,' says she, gasping for air. 'I can hardly breathe.'

'It'll be okay in a few moments. Okay, fingers off the trigger and hold on. We've got you.'

A dull thud on the bulkhead: Rose stands, fighting to suck in the rushing air, the knuckles of one hand white from gripping the rim of the helmet, the knuckles of the other white from gripping the lance, finger still firmly on the trigger. The other woman takes a half-step toward the three suits.

'Finish that step and I'll cut you.'

'But I need a suit, damn you.'

'You've got a suit... Lenny's. Move again I'll kill you. Take your chances.'

The airlock spins open. A fine mist appears and the cabin fills in a chaos of decaying debris. Despite the rush of air Rose manages to stay on her feet, she staggers but remains upright, the lance not leaving its target. The second woman slumps to the floor in a faint.

The hatch now fully opens and two suited crewmen enter. Rose immediately lowers the lance and moves to place the two men between herself and the crouching second woman, who now staggers up and stands holding her head in her hands and looking very sick. Rose raises the lance again and screams at the two crewmen,

'Get me out of here before I cut her... Get me out.'

The first SBS crewman, Cameron, takes the lance from Rose. The other crewman leads her through the hatch. The moment Rose leaves the ship she breaks down and sobs uncontrollably. The airlock closes behind her, leaving Cameron alone with the second woman. With the lance still held firmly in one hand he twists off his helmet with the other. It comes away with a hiss of air to reveal a rugged, good-looking thickset man with cropped hair and *Captain Kirk* side-burns. The woman recognises him.

'Cameron, isn't it?'

'What's it to ya? Just move. Get into the coffer. I got to lock you in, Ro–' he stops mid-word.

'Com'on, Cameron, you know me. I'm Rose, Rose Hawkins. You nearly said it. Com'on you creep.'

'Damn it, you're not Rose. Just move.'

'You know me, Cameron. You made a pass at me once, you bum. I'll tell you what your line was when you grabbed my butt, "Hi, how are *you*, sweet-chips?" Remember, you goddam pervert? Com'on, I'm in pain, I need help.'

She takes an ungainly step towards him. Cameron looks momentarily into her soulless eyes then hurls her away in disgust. 'Get away from me.' She falls to her knees onto the debris-covered deck. 'I made a hit on Rose, not *you* – you make my goddamn hair stand on end. I wouldn't touch you with a cattle-goad. Now get in.'

She staggers up from the deck, speaks as she brushes off the putrefying mess from her suit. 'Well, you've got to call me something. How about something clever, like… 'Rosette', you stupid jerk?'

'Yeah, and fuck you too. Just get in.'

'How in God's name did a creep like you ever make crew?'

'God's name? Satan's more like. Now get in.' He takes another step towards her and seems about to grab her.

'Back off. Don't touch me you creep. I'm sick, I'm in pain.' She starts to climb into the remaining coffer. 'I'll go quietly, but I need something for the pain… my whole body is on fire – I'm begging.

'Sorry. When we get to EarthLab. There's nothing I can do here.'

She lays down in the coffer. It closes. Cameron stands

over the transparent lid and studies her. They stare eye-to-eye for what seems an age. She eventually looks away. Cameron puts the lance down and secures the coffer.

He staggers slightly under the tug of the SBS as it begins to tow Junairo to EarthlabOne.

CHAPTER TWO

Edinburgh University shines out as a redbrick oasis of colour, in an otherwise gloomy grey-stone city. It is raining and almost dark in spite of being just mid-afternoon on a miserable winter's day. Inside the neglected Victorian building is a seated receptionist attending two standing men. She smiles and holds up her hand, 'please wait.' She picks up her telephone and speaks.

'Mr. Mandrake, there are two...' she eyes the two athletic-looking men, '...gentlemen to see you.'

In his office, two floors up, Harry jolts bolt upright at his desk. He speaks nervously into the phone. *'Gentlemen* you say? What are they like?'

The receptionist looks up at the two enormous hunks again. Embarrassed to answer, she whispers into the mouthpiece, 'They are Americans.'

'So they're bloody 'Americans'... they are human I take it? What the hell are they like, for Christ sake?' He yells into the phone cradled under his chin as he nervously and energetically gathers his papers, 'Are they big? Are they mean? Are they wearing black hats or white hats? Jesus Christ girl, use your bloody loaf.'

The receptionist, miffed at his ratty attitude, retorts, 'Sod off. I'm sending them up.' She bangs the phone down.

'No no no. Oh, Christ, stupid bitch,' weeps Harry, letting the phone drop from his chin. Without bothering to replace it on the receiver he dashes about his office with inspired panic grabbing, coat, hat, and diary, and as many papers as he can carry. Then he's off into the corridor, avoiding the elevator, leaping down the emergency stairs three at a time and crashes out through the doors into the square.

In the college quadrangle, Harry runs for dear life through the drizzling rain. One of the Americans steps in front of him.

'Henry Mandrake?' He challenges.

Harry stops dead, almost falling into the big man's arms. 'Christ almighty. You scared the hell out of me. – Mandrake? No. Why? What do you want with him?'

'Calm down, buddy.'

'Mandrake left ages ago,' says Harry dismissively. 'Can I give him a message? I'm in a deuce of a hurry… freezing cold… papers getting soaked, do you see?'

'Look… I know you're Henry Mandrake,' says the American, 'I got your photograph. Whatever you're thinking, you are wrong. I'm from the Carnegie Space Agency PLC. We want to talk to you… to your advantage… money. Savvy?'

Harry blinks nervously at the American, hardly understanding. 'Money, you say? Money. What the Devil do you take me for?'

'Let me introduce myself. I'm Rex and this is Hamish.' The other American now joins them.

Harry, feeling the immediate danger having passed, begins to take interest. 'Right, right you are… Rex. Money, you say?'

'A great deal of money,' winks Rex.

'I know a place – Oh… you have expenses? Of course you do… silly me. Let's get out of the rain… I'll lead the way. The name is Henry, but I prefer Harry, *Hal* to my friends. My father named me after Monmouth Harry.'

They look bewildered. Harry continues. 'Henry the Fifth, part one... Shakespeare... Azincourt, Harfleur, *"once more into my breeches,"* do you see?' Still no response, Harry shrugs in dismay, 'Dear God. – I take it you have transport? Follow me. Mine's the vintage Roller.'

The granite façade of Gentlemen's Club, *The Deacon Brodie,* glisters wetly out of the Edinburgh mist. Harry jauntily enters the foyer of the oak-panelled mansion – a relic of Victorian grandeur – followed by Rex and Hamish. He beckons a waiter.

'Right, Robbie... table for four... a lady will probably join us later. The Deacon's Cabin, I think. Some drinks for starters and a bit of privacy.'

Robbie looks apologetically to Harry and speaks to him, aside from his guests, 'I must remind you, Mr. M, your account. Sorry, but it's from the top. Nothing I can do.'

'No worries, old sport,' sings out Harry. 'The good old Yankee dollar.' He turns to Rex. 'Look here, Rex, you'll have to leave your plastic with Robbie, do you see? Bit embarrassing... you've no idea how little a lecturer's pay is. I'm just going to make a phone call. You follow Robbie.'

Rex fixes Harry a sour stare as he hands over his *American Express* card to Robbie's outstretched hand, then he and Hamish follow the waiter's lead into the annex – a glass panelled cubicle in the centre of the dining room.

Harry makes his way to the phone booth in the foyer. Inside, he looks up a telephone number in his little notebook, dials, and waits as the phone connects. 'Ha ha. Guess who? Yeeees, Sweetie, righ–' The phone goes dead.

Unruffled, he looks up another number and dials. 'Ha ha. Guess who? Right Sweetie, 'Harry.' Going to buy you din– Steady on, of course I'm paying. Sor– let me get a word in. Sorry about that. I'll make it up to you. You know how, ha ha. And I've got a little pressie for you,' he twists a pretty necklace through his fingers as he continues, 'been in the family for yonks... sparkly. What say, Sweetie? ... You will. Good girl. Get a cab to my club... ask Robbie to put it on the bill. Oh, and bring your overnight bag. See you in an hour, *mwar, mwar, mwar.*' He kisses down the phone then tosses the hand-piece back onto the receiver with rakish style, letting out his cry of triumph as it lands neatly into place: 'Bloody bingo.'

Rex and Hamish sit patiently waiting, arms folded and staring indifferently into space. They are seated at an elegant antique dining table in a small, casement-windowed annex set in the centre of the main dining room, the Deacon's Cabin – a room within a room.

They come to attention as Harry, now in top form, opens the door, speaking as he enters, 'Right, Rex, Hamish, what do you want in the trough?' They look bemused. Harry expands, 'What's it to be... eats, drinkies?'

'Just mineral water for us,' says Rex, dismissively.

'Slice of lemon with mine,' adds Hamish. Harry smiles and takes his seat at the far end. Robbie knocks on the door and enters. He raises his eyes to Harry, prompting the order.

'Two mineral waters, one with a slice of lemon,' says Harry, 'and I'll have a triple G&T... oh, and a bottle of Moet and one glass... and whatever you're having, Robbie... on the old billy-do.'

'Jessus.' Rex rolls his eyes to the ceiling in amazement as Robbie goes about his business. He then shakes his head wearily. 'Okay, Harry, to business: I take it you've heard of the USS Junairo, the Mars space shot?'

'The starship that caught both titties in the mangle?' says Harry with a liberal hint of sarcasm.'

'Yes... literally. Damn good analogy – four 'titties', to be precise. Now, I want you to realise something, Henry–'

'Harry, please.'

Harry... pretty Harry... what I'm about to give you is privileged information. Do you want me to go on?' Harry remains silent. Rex fixes him with an uncomfortably long stare. At length, he continues. 'Because if I do go on, and you then refuse to help, things could get nasty. So I repeat, do I go on?'

Harry mulls the unquantified proposition, of which he is extremely wary but he's also extremely short of money and things are already 'nasty'... namely *Radcliff,* his thoughts on whom quickly resolve the quandary. 'How nasty?'

'Nasty, nasty.'

'Humm... That nasty? Go on.'

'Okay Harry, but first I need to know more about you and your late uncle, Barnaby Mandrake, Lord Melrose. How well did you know him?'

'Very well, and I hope to know him very well again. He's only 'missing, presumed dead'... I'm hoping he'll turn up.'

There is pregnant silence for a few moments. Rex picks up his drink and takes a swallow, his eyes not leaving Harry.

'He's been gone eight years, Harry,' says Hamish, now taking over. 'That's a long time to be missing. Rumor has it he blew himself to kingdom-come with a rocket engine he was developing... sounds about right to me.'

'As I say, 'missing'.'

'So, with him being 'officially deceased' you inherit everything... the entire estate, I understand?'

Harry shrugs, 'Well, you understand wrongly, old sport. Do you imagine I'd still be wintering here if I had monies?'

'We were told–'

'You're partly right – Encumbrances.'

Hamish puts his hand to his jaw as he considers.

Rex gives a shrug and finishes the last of his drink. 'Encumbrances, you say? Explain.'

'Yes, old man... encumbrances. To inherit I have to pass my Master's degree, marry and produce an heir... *Marry.* I bloodywell ask you... do I look the marrying type? – So, I stay here under sufferance, as chief lecturer on PXL.'

'P X L?' queries Rex.

'Possibilities of Extraterrestrial Life,' explains Harry. 'I'm quite a success, much to their annoyance, particularly with the ladies. Ha ha. Do you see?'

'No, I don't see. 'Under sufferance?' 'To their annoyance?'

Harry, realising that Rex is troubled, quickly adjusts his story. 'Yes... well, you must understand... the old man, my illustrious uncle, injected a massive wedge of monies into the founding of the academy, creating the new PXL section of the science faculty. With added proviso that I have top place as lecturer and custodian of his papers, cetera-cetera, blardy-blar... Gets up their noses I suppose.'

'Gets up their noses? What in hell does that mean?'

'They don't bloody like it, matey... Jealousy. Nothing

they can do about it.'

'The 1950's *Mandrake Experiment,* Harry...' says Hamish, pushing the conversation on, '...One man eaten alive in his flying suit the other left a homicidal maniac?'

'Good Lord. I'm surprised you've even heard of that, let alone believe it. Most people don't give it a dot of credence.'

'We have an open mind, Harry,' says Rex, offering a face of reason. 'We can't afford to overlook anything. It's the only account of PXL, as you call it, that is anyway similar to our own, shall we say, 'dilemma'.'

'And my God...' says Hamish, '... is it similar.'

'Your government won't release all the data,' continues Rex. 'Some of it is still in the archives "under embargo", so they say. They won't be moved. I smell a rat.'

'Yeah, a great big dirty rat.' Hamish chips in.

'Dirty?' says Harry, seemingly puzzled.

'Yes, *very* dirty,' says Rex, raising his voice. 'We think that rocket engine Lord Melrose was working on was nuclear, and that it was designed in Nazi Germany by a man named Ulam. The Russians captured it after WWII. They were scared, or didn't have the expertise, to test it. You Brits traded the Rolls Royce jet engine and the turbo-jet technology for it. You know what I'm saying here?' He stops and studies Harry, waiting to see if he comprehends what he's implying.

'Of course, I know what you're saying... you are, after all, speaking bloody English... of a kind. Go on.'

'The Russians used that jet engine in the Mig 15, smart-ass. We lost a lot of boys in Korea because of it.'

'What total rubbish.'

'That's a goddam fact, buddy,' growls Rex. 'Your crazy uncle used his wartime connections to broker that deal. With what he got in return for that jet technology, he was able to send a manned atomic fucking ballista; a goddam ironclad battleship space-shot, to Mars.'

Harry feigns aghast. 'You're thinking of the 1948 Project Orion. That was a crazy American thing. But that came to nothing. I mean to say, dumping little atom bombs out the back of a spaceship and catching the blasts with a pusher-plate... absolute lunacy.'

'It was nothing to do with Project Orion,' yells Hamish, exasperated. 'Your crazy uncle's *Mandrake Project* was the lunacy. He enlisted a British submarine-expert to build a spacecraft. That engine was so powerful the vessel rounded Mars and crashed back to Earth, leaving a trail of poisonous radiation. It was under full power, both ways.'

'Never. You don't seriously believe that, do you?'

'Yes, we, fucking do, believe that. I further believe that reactor leaked and contaminated the crew. The radiation was so intense it disintegrated one man and sent the other crazy. That, you insufferable Limey creep, is exactly what I think.'

'Now, steady on,' says Harry deeply offended, turning to Rex for support, 'I mean to say, do you chaps have to swear?'

Rex shakes his head. 'Sorry, Harry, that's what we all think. So...?'

'Yes, well, that's as maybe, Rex, old sport, but–'

'A great deal of money, Harry... So...?'

Harry considers. He's got two options: one, do what they ask, or two, do what the Gorbals nightmare asks. 'Okay... I've got the official edited version of the flight, and transcript of all data plus various specimens.'

'Just the transcript, Harry?' says Rex, unimpressed.

Harry conceders a moment, then decides to juice up his

story a little. 'Look here, old man, I've got exclusive access to archive material... and–'

'Just access?' says Hamish.

'I'll let you boys into a little secret,' says Harry tapping his nose and lying through his teeth, 'I've got the original film. They don't know it yet, silly buggers.'

Hamish shrugs, 'How come?'

'I'll tell you, how come?' Says Harry, desperate to think of something that will satisfy. 'I'll tell you how come... ah, yes, right... well, you see, there's a hell of a lot of material there of which I'm not supposed to have access to, let alone take out of the place. It's all kept together and I'm given what I ask for by an ever-present attendant. He decides what I can and cannot see. So I go there, regularly. They think it's for material for my lectures–'

Hamish shrugs impatiently, 'So, how come you got the original film?'

'I'm coming to that. So I snoop, photocopy and... purloin, steal the originals. But here's the best, I took my edited copy film there for comparison with the master. The old twot of a projectionist actually handed me back the original un-edited negative by mistake. It was still in its original spool. Anyway, it's mine isn't it? – It's my bloody living for God sake.'

Rex looks reasonably impressed. He continues in a slightly friendlier voice, 'Okay Harry, I'll go on, but you've been warned. As you so rightly say, the Junairo failed... a bug got in the works, not radiation or gamma poisoning. Two men, Captain Leonard Cowen, Allen Fitzgerald, and a woman, Rose Hawkins, blasted off from EarthlabOne in August... perfect. Closed on Mars in October... perfect. They took some mind-blowing pictures, you probably saw some of them in the media?'

'Yes, very impressive.'

'Then the trouble started. Junairo went behind the planet and... zilch, nothing. They were not to land on the planet's surface, just to soft-land survival modules – provisions and fuel – at various proposed sites for future planned landings.'

'What,' exclaims Harry, flabbergasted, 'you mean they went all that way and were not going to land?'

'That's right,' says Hamish. 'There were to be two more journeys before the actual landing... the stay would be for three months. Have you any idea how much support material that would take?'

'Yes I have, actually. But the cost... what amount are we talking?'

Rex studies Harry. 'I don't think you need to know that, Harry. The goddam President don't even know that. He thinks the money is for some cockamamie star-wars project of his, defense against UFO's – oh yes, he claims he's actually seen them – Elvis is alive an' living in fuckin' Disneyland DC.'

"Hey,' says Hamish, 'show some respect for your President.'

'Respect. I had respect for that guy ever since I saw him in that Hemingway film, *The Killers*. Anyway, let him think what he likes, it keeps him off our backs. So... the Junairo Starship is permanently in space and is fuelled and refuelled from EarthlabOne. It is reusable, endless times. The cost is negligible, that's all you need to know.'

Harry shrugs, duly impressed. 'So, what went wrong?'

'We don't know, we lost all contact. We didn't pick it up again until it was almost on top of us. Our screens were down – jammed with some weird form of static. When we got to them, the men were... gone.' He stops and puts his hands over his eyes, wanting to blot out the offending conundrum.

'Go on, go on.' says Harry, eager eyes, and willing ear.

Rex shrugs. 'Christ, it sounds crazier every time I hear it. Damn it, they were turned to ash and slime. Like they glimpsed the goddam Medusa or something, just the woman was left. But–'

'Sounds very similar, Rex, old man,' says Harry, 'Very similar in deed.'

'That's not all, Harry. There are *two* of them.'

'Two? I don't understand.'

Hamish, now over his malaise, takes up the story. 'The same. Two women exactly the same: Rose Hawkins and an exact double, a doppelganger.'

Rex bangs his hand down on the table emphasising the dramatic, implausible explanation. He studies Harry's expression. Harry nods uncomfortably. Rex leans very close into his face, deliberately invading his personal space. 'You're the only person outside the Agency in on this, Harry. Let me explain something to you – whereas the Agency is a private concern, it has, in many areas, been in direct collusion with the US government's so-called Star-wars Space programme, EarthlabOne. We are allowed certain, shall we say, latitudes... because it's in everyone's interest.'

'We can do things, Harry,' says Hamish taking up the story, 'We, unlike the US Government whose space-exploration budget was put on hold for the duration of the Vietnam War, are not encumbered with military restriction. We've done a secret deal with them. They have their voters to answer to, we have our shareholders.' He stares into Harry's eyes for an uncomfortably long time.

Harry doesn't look away.

'Go on, old man.'

'We have a private financial consortium,' says Rex, continuing the tale, 'You Limies, the Japs, French and the Germans, even the goddam Ruskies are on the payroll – no government can do it alone. Conglomerate money is the way to the stars, private enterprise – we take the money and don't ask no questions, from all comers, both sides of the law... we're *connected* you might say.' He gives a menacing stare as he continues. 'So you see, Harry, if anything of this leaks out... if anybody jeopardises this project... if we're grounded... very nasty.' He winks again. Harry nods back uncomfortably. Rex studies him for an overlong time, then winks again.

'Will you *stop* bloodywell winking at me.' growls Harry, infuriated. 'I understand, Rex, I understand. Dear God.'

'I don't think you do, Harry. It's like you Brits with your monarchy. You give them enough wealth and power to make them, supposedly, incorruptible. Then, if one does fuck up, the chop, literally. I like you, Harry, don't fuck up.'

'It's like this, Harry,' says Hamish, now playing good cop, 'we need to protect ourselves, our people. The Agency can't risk a crew again until it has that protection. We understand your uncle had developed a shield. We want you to release all your data to the Agency for six months. Name your price.'

Harry sits back, picks up his gin-and-tonic and raises it as if in toast. 'Hamish, old love, I'll do better than that,' he drinks it off in one swallow. 'You get the data, and you get me into the bargain. We'll consolidate my fee later.'

'You'll come along, Harry?' says Rex, seemingly

aghast, 'To what purpose?'

'Use your loaf, Rex. I've studied this since I was a child, day in day out. I couldn't get away from it, bloody thing haunts me.'

'So...?'

'So I know things, theories, hunches, things you won't find in any data.'

'How come?'

'My uncle spoke of nothing else. He knew this would happen, he knew it was out there waiting. That's why he developed a shield.' Harry stares at Rex, then to Hamish, and then back to Rex, wallowing in their rekindled interest.

Hamish pinches his eyes shut in renewed exasperation. 'Go on...'

'That's also why he set me up here, at the University. He wasn't sure he'd be around. And anyway, I'm custodian of his works, I must insist. I owe him that much.'

Rex studies Harry, staring at him, again for an uncomfortably long time. He suddenly leaps across the table, 'Good. Excellent. You're in.'

'Christ!' gasps Harry.

Rex grabs Harry's hand. 'When can you leave?' Harry tries to pull his hand away, Rex holds on to it. 'When, Harry?'

'Ready when you are, squire,' says he, as the spectre of Radcliff subliminally flashes into his mind's eye, 'Sooner the bloody better.'

Rex releases Harry's hand and leans back in his chair seemingly contented. Harry looks nervously to Hamish. Rex looks back and smiles. 'In two days, that's when we go back. Midnight Saturday. We've booked your flight already... an expedient. We expect you to give a complete debrief on your 50's Mandrake Experiment on Monday. What do you say, Harry?'

'Call me Hal, old sport,' says Harry, now sporting a winning smile. 'And I say, done deal. Now, to show your good faith I'd like a small advance. Oh, and settle my outstanding account here when you pay the bill. I have to leave my affairs in order... do you see?' He gives a nervous shudder as he again considers the Radcliff debacle, 'I have a few other things to put in order,' he winks at Rex, 'So, how say, old man?'

Rex nods 'yes.' Hamish rolls his eyes. Harry opens the champagne, pours and drinks off a huge glass, refills and drinks another.

A tap-tap on the door. Robbie sticks his head in. 'Excuse me, Mr. Mandrake, the lady has arrived. Shall I show her in?'

Harry gestures for Robbie to wait. He turns to Rex and smiles. 'I take it we've finished business?'

'Yes. I've no need to remind you of the gravity of this discussion, Harry?'

'*Hal.* No need. Don't concern yourself, Rex old man, no need.'

'Remember, Hal,' says Hamish with menace in his voice, 'what we said – cabbages and kings – the chop.'

Harry winks. 'Lips sealed,' says he, 'wild horses... cetera-cetera, blardy-bla.' He turns to Robbie. 'Send the lady in, old luv... Oh, and bring another bottle of Moet and another glass. And Robbie, add my entire account to this bill.

Robbie nods and walks off.

Rex hands Harry an envelope. 'Your ticket and expenses plus your advance: half in 'Yankee dollar', half in Sterling. Again, anticipated. We're leaving now. See you at Heathrow, VIP lounge. Bring everything.'

'I take it it's first-class.'

'Of course.' says Hamish, slightly aggravated.

'Right you are, then. I'm impressed.'

Robbie returns, ushering an attractive young woman. She eyes the two tall Americans with flirtatious smiles.

Rex and Hamish nod hello/goodbye then follow Robbie out of the annex without further word.

Harry calls to Robbie, 'Tot up the bill, old luv. Wendy and I will dine and stay the night. Oh, and breakfast, you know the sort of thing... stick it on the bill, the nice American gentleman is paying.' Harry winks again at Rex, standing just outside the door. Rex does not reciprocate.

Robbie leads the two Americans to the desk. They stand idle while Robbie tots up the bill. He hands it to Rex.

'Son-of-a-bitch!' gasps Rex, for the first time losing his cool, 'Look at this. That's goddam *pounds*, not dollars.' he shoves it under Hamish's nose, 'Jeeesus.'

Hamish shrugs, 'Pay it.'

Rex rolls his eyes, 'Son-of-a-bitch.'

CHAPTER THREE

Alfred Watkins, a dapper 70-year-old Englishman, whistles tunelessly as he busies himself, polishing and dusting the palatial apartment. He carefully avoids the myriad of modern and vintage electronic equipment: satellite, DVD, CD, flashing modems, video, fax and even an ancient telex, plus every modern computer IT communication peripheral. This is Harry's abode, everything remote-controlled, from the curtains to the dishwasher—remote control hand-units are scattered in every room of the apartment.

Outside in the hallway, Harry fiddles clumsily with his keys, his hands being full of packages. He opens the door, enters and smiles as he sees Alfred, pottering. 'Alfie, old mate, leave that... Good news.'

'I'll put the kettle on... Sir?' says Alfred, aloof, in a common-as-muck London accent.

'Oh Lor,' sighs Harry, wearily. 'We're not back to 'Sir', are we? I'm sorry I didn't tell you, but you know... a lady.'

'It may interest *Sir* to know, that I too have a lady, and she bloodywell likes to see me from time to time. Last night's dinner is in the bin. Do I put the kettle on or wot?'

Harry lays down his packages and flops into a big soft Chesterfield, sighs again and closes his eyes. 'Alfie, be a love, don't be cross. I've got good news and bad news: I've got to go away for some time.'

'Away? Where?'

'You don't need to know 'where' or how long, I'll keep in touch. Just keep everything going – don't forget to water my mandrakes, not just the potted ones, remember the outside ones, and keep the frost off.'

'Hey, I'm not your bloody gardener.'

'Move your lady in for a while if you like. Now the good news.'

'Oh... I thought that was the good news'

'Yes, that is funny. I'll laugh later when I've got less time. Now, I can pay all your back pay, your holiday pay and a month's advance wages plus a little bonus... how's that, old mate? When Harry Mandrake is in clover, everyone's in clover.'

'Yeah, and when Harry Mandrake's in the shit, everyone's in the shit – We're not in the shit are we, Harry?'

'How dare you. No, I'm not in any trouble, but you will be if you don't make that bloody tea. Now, leave me alone to make some calls. Pack my bags or something... not much... I'm on expenses.'

Alfred shrugs and leaves the room. Harry rests for a few moments, then rises, picks up the telephone, trailing its long wire as he steps out onto the veranda overlooking a panoramic view of Edinburgh castle. He picks a new mandrake blossom from his window box and replaces his buttonhole, gives it a sniff then dials. The call connects:

'Hello, do you recognize my voice – *don't* say my name – just say yes, if you do? ... Right, now listen. I'll pay you in full, Saturday... No, I can't pay before. It has to be Saturday. I've got to go to London to get the money.' He stops and listens for a few moments. 'Good. We'll meet at the old university offices in Wells Street: write this down, I don't want any mistakes—you can write, I take it?

'... Sorry. Yes, I'll leave the door unlocked. Take the lift to the third floor, the <u>third</u> floor, got that? And wait for me there, you'll find it. Half past eight in the evening, not a moment before. Got all that? And try not to be conspicuous. ... What I mean is try not to let anyone see you. With a bit of luck, they'll all be gone by then, so you won't embarrass any... It was a... it was a joke, for Christ sake. I'm sorry. ... Yes in full... yes... Yes. Goodbye.'

Alfred brings in the tea tray. Harry emerges from the veranda sporting a boyish grin. He triumphantly punches his hand into the air, then tosses the hand-piece of the phone over his shoulder, spins around and catches it onto the receiver, 'Bloody bingo.'

'Wot you been up to, I know that faffin yell... that means trouble?'

Harry ignores Alfred's rhetorical question. He takes his tea and places the telephone on the tray. 'Get London on the phone, Alfie: Tell them I'll be coming to check my uncle's documents. I'll need the cine-projector and the video equipment... first thing in the morning. If they quibble tell them I'll also settle my arrears.'

'Bloody hell. You feeling okay?'

'I'll travel to London tonight. Book a flight. Oh, and also settle my account at Ratner's, I need to pick up a few sparklies while I'm there.'

'Load of sparkly crap, you mean.'

Harry shrugs and flops back into the sofa and drinks his tea, then surrounds himself in cushions and dozes, leaving Alfred to make the arrangements.

After grudgingly completing his chores, Alfred wakes Harry with a sharp shriek into his ear, 'DONE, MASTER.' Harry startles, Alfred smiles, and continues,

'Your flight is at seven. London didn't mention your arrears, so nor did I. The Ratner's account, I said it would be paid in full when you call tomorrow. Any other whim I can satisfy... Sire?'

'Excellent. For this, I award you your freedom... go and be fruitful. Now, bugger off. I'm going to get some kip.'

Harry pours another cup of tea then disappears into his bedroom.

CHAPTER FOUR

London, unsurprisingly, is raining. Harry spends the cab journey from the airport, sleeping. Now, as he enters the white stone-pillared building, housing the City of London Archive, his head is aching. He takes the lift to the fourth floor to the reception, and smiles at the attractive woman receptionist.

She recognises him and smiles back, 'Mr. Mandrake, how nice, we haven't seen you for ages. It's all ready, Mr. Hamon will be your attendant today.'

'Don't need him this time, Brenda, just a quickie… an in and out visit.' He takes out his buttonhole and places it on her desk. 'For you.'

She picks the violet blossom up and sniffs it, smiles and hands it back, 'You better put this back, you don't look right without it…Sorry, Harry, you know the rules.' She stands and ushers Harry into a small, badly lit room. A scrawny, middle-aged Mr. Hamon enters, awkwardly carrying a large metal box under his arm. He nods his head to Brenda as she leaves, then he turns and scowls at Harry fiddling with his lapel.

Harry eyes the box, 'Ha ha. The old Ark of the Covenant.'

'None of your lip, young Mandrake, I haven't got all day to waste on your rubbish. What do you want first? Make haste.'

Harry shuts the door and, unseen by Hamon, takes a small glass phial from his pocket, passes it behind his back and drips a few droplets of fluid onto the floor behind him.

'Do apologise, Hamon, I had the most fearful vindaloo last night... didn't sit too well with the old champers.'

Hamon looks puzzled, then shakes his head dismissively and places the box, still locked, onto the table. He stands back and fiddles for his key, then reacts to a rising smell. He gives Harry a very suspicious, damning look.

'The old Derby Kelly, old luv... bit upset,' says Harry, sheepishly, 'Do you see?'

'Yes, unfortunately, I do see. And if you don't mind I'll step outside for a minute.'

The moment Hamon is gone, Harry picks at the lock of the box with a bunch of wire keys. It opens and rifles the interior, taking two ancient-looking spools of film, a bunch of papers, and two sealed bottles. He replaces them with his own, look-alike spools, papers and bottles. He closes the box, locks it, and then backs away.

After a moment Hamon returns, sniffing the air and eyeing Harry accusingly, 'Are you going to be long, Mandrake, I have other, *important*, things to do?'

Harry, with one hand on his stomach the other hand behind his back, drips a few more drops of the ghastly fluid from the glass file onto the floor.

'Don't think I'd better go on, Hamon... don't feel at all well... bit stuffy in here... the old collie-wobbles, I do apologise. Another day. Perhaps tomorrow.'

Hamon, catching the smell again, grabs the box and starts to hurry out. Harry just manages to flick a few drops of the evil fluid onto the back of the departing coattail. He can hardly contain his mirth as Hamon hurries out, past Brenda, muttering to himself. After a few moments Harry walks out, briefcase in hand, up to Brenda. By the look on her face, she has obviously smelled Hamon on his way past.

'I think Hamon's getting too old for this job, Brenda,' says Harry, smirking, 'See you next time... Goodbye.'

An assortment of cheap, costume-jewellery pass under Harry's meticulous eye as he makes his choices. The attractive Ratner's shop assistant, boxes and wraps, and then charges the whole to his account – no mention of the unpaid bill, none offered. She smiles dutifully and hands the bag and a docket. Harry signs, takes the bag and smiles back, winks and leaves.

On the Edinburgh-bound aircraft, Harry fumbles through the pages he has stolen. After a few moments, he puts them away and looks out of the aircraft window. London, far below, is still raining. Mercifully the dismal metropolis disappears as the aircraft gains altitude, heralding a brief glorious burst of sunlight. Then cloud again, the sky towards Scotland looking dark and ominous. Harry is bored, his attention easily diverts to the attractive hostess. They exchange smiles as she brings him his drink. He flirts.

Further along the aircraft a small, wavy-haired man sits furtively watching. The man studies Harry intently, ducking down and hiding his face surreptitiously every time Harry looks up or passes by to the toilet.

Moving past the wavy-haired man to the last window, out through the tensile glass and into the cloudy thin air, then on into the stratosphere, and onwards and upwards, into the total blackness of space, closing on EarthlabOne. The space station emerges from a dot of light. Big shapeless; a jumble of cylinders, pipes, and gantries, plus four tethered SBS hanging from long squid-like tentacles. Passing through these obstacles and into the heart of the craft, the internal structure gives way to a network of passageways. Along the main passage three men escort a woman. She is wearing a restraint jacket and has a pained, sickly pallor. In claustrophobic manufactured quarter-Earth gravity, they lope with effortless gait.

At the middle of the passageway the group stop and enter the main control-deck, a circular gallery where Rose, Major and Cameron, and four technicians await. Major holds a computer printout in his hand. As the party enters, he attempts to read it to the restrained woman. He hesitates and turns to Rose.

'What the hell do I call her?' Rose stares daggers back. Major looks at the rest of the group.

'Call her Rosette,' offers Cameron, half joking.

'You kidding me?' says Major.

'No, I'm not – if you have a better idea...'

Major shrugs, 'How about, Rose Two?'

'Christ's sake,' yells Rose, slamming the clipboard she is holding down onto the desk. The noise echoes around the room. She snatches the report out of Major's hand. 'Can we get damn well on with it?'

'Okay Rose, calm down.'

Rose gives Major a damming look, then turns her back on him and speaks directly to the woman, 'Whoever, whatever you are... Rosette... you have no spleen and a malformed lymph node disorder. After vigorous tests, we've identified a rogue cell structure. We've tried to isolate it, but you need expert consultation. Other than that you are an exact duplicate of me, except for slight DNA and RNA anomalies.'

'So...?' shrugs Rosette, indifferently.

'So, that's it. We're taking you down – we can't do any more for you up here. If I had my way I'd dam'well jettison you right here in space.' Rose takes a step towards the woman, glaring into her eyes with malice.

Major steps between them, 'That'll do, Rose. This is difficult enough without that.'

Rosette gives a chilling half-smile, 'Thank you, Major.'

Major ignores her, and speaks again to Rose. 'How are we doing in England? Did you manage to contact the boys? When will they be back?'

Rose breaks from her icy stare. 'What? ... Oh, a couple of days. Rex and Hamish will be here early Sunday morning. The Brits won't play, they're sticking to the embargo on all Mandrake data, but I think we've overcome it. It's imperative we get that material. As to Henry Mandrake, we need him desperately.'

'I want them back earlier. Is that a problem?'

Rose shakes her head, 'Yeah, I think so. But I'll try.' Still angry she turns and walks through to the transit bay.

Major and the rest of the party follow through an airlock marked 'SBS ORION', for the journey back to Earth. A few minutes later the shuttle detaches and slowly eases away from the main structure. In a controlled gas-jet glide it enters Earthbound trajectory and on into thin atmosphere, then dense cloud, and then through to wispy vapour.

At the same time, halfway across the world, Harry's Aircraft is approaching Edinburgh Airport.

An hour after landing Harry enters his apartment, to Alfred's icy greeting. Pleasantries and unpleasantries exchange, then Alfred grudgingly makes the tea. Harry moves to his study and busies himself at his desk with bits of gold wire and electronic equipment. After completing two electric circuit boards, he takes a small metal box from his desk and carefully opens it. It's full of inch-long, coral-coloured pellets wrapped in cotton wool. He takes four and carefully links two to each of the two devices. Alfred brings in the tray of tea, pours two cups and sits alongside Harry.

'Wot the 'ell are you up to, faffin about with wires and bits of Tom-bleedin'-foolery?'

Harry gives a puzzled look. 'Tomfoolery?'

'Jewellery.' qualifies Alfred, curtly.

'It's not jewellery.'

'Well it looks like faffin jewellery, bloody beads an' gold wires. Anyway, wot you making?'

'Guess who I saw today, Alfie?' says Harry, ignoring the question.

'Surprise me.'

'Our little curly-haired friend... he was on the plane.'

'Did he see you?'

'Of course he saw me, he's bloodywell following me isn't he for God's sake.'

'Okay, Mr. Smart-arse... only asking. Did he see you see *him*, that's wot I meant?'

'Sorry. No, he didn't. I haven't seen him for ages. Three Americans in two days... puts the price up, wouldn't you say?'

'How'd you find out Curly's a Yank?'

'Heard him ask the hostess for a *Scutch.*'

'And you still don't know who he is? '

'Nope.'

'And you're still not worried?'

'Nope. I have bigger things to worry about.'

Harry hands Alfred a package marked Ratner's. 'Oh,

and stick these in the old Scrubs ammonia bottle, Alfie... age them up a tad before I go.'

'I don't like it, Harry. I promised your uncle I'd watch out for you. I can't do that if you won't tell me where you'll be, can I?'

'Where do you think the old man is, Alfie? Do you think he's still alive?'

'No, unfortunately, I do not. How many more times? I'd know if he was alive. For my money, he's definitely brown-bread. But wherever he is, alive or dead, you can bet there's bloody trouble, where there's a Mandrake there's always bloody trouble.'

'Give it a rest, Alfie. I'm going to have an hour's kip, as you call it, then I'm off.'

Alfred rolls his eyes and walks off into the kitchen.

CHAPTER FIVE

The antiquated annex of Edinburgh University offers itself uncompromisingly to a dull, forbidding evening gloom – it is still drizzling. Harry turns his jacket collar up as he walks quickly from the car park. He unlocks the big green doors, enters then locks them again behind him. In the deserted reception area, he summons the elevator and waits. The old-fashioned iron-cage lift rattles to a halt. Harry enters, closes the door and pushes the button for the third floor. He meticulously times the assent and makes various notes on the back of his hand with his ballpoint pen. This operation he repeats two more times then returns to the ground floor. He opens the control box and removes a large section of wiring and affixes one of his two coral-coloured-pellet devices. The other device he fixes into the trap door in the roof, he then replaces the covers and leaves the lift, closing the cage door behind him. As the ancient doors rattled closed he looks at his watch, his face expressing a smug, 'everything is going to plan' smile. He turns on the reception light, which automatically turns on the lift light, then he unlocks the big green doors and peeks out into the now-misty twilight. A car's headlights approach. He quickly ducks back inside, leaving the door ajar, and hides in the shadows of the stairwell. A few moments pass.

Radcliff enters, twitching, neck jerking and looking villainously from side-to-side. He walks towards the lift, oblivious to Harry secreted in his hideaway. As he swaggers up to the lift door he has not a care, worry or fear in the world. Regardless he takes a last instinctive look over his shoulder, twitches one last involuntary twitch, and enters.

Harry studies his watch. He takes a small remote-control unit from his pocket and, from the darkness, points it at the lift. He has the glint of murderous-intent in his eyes as the lift starts its clattering journey upwards. Harry again looks nervously at his watch, then to the lift, then to his watch, and back to the lift. He aims the remote, his finger ready on the button. A few more moments pass, then he presses. A dull, muffled explosion followed by the sound of searing metal. The lift judders and stops dead, fused exactly where intended between floors. The escape hatch in the ceiling is also seared and welded shut. A few seconds of silence pass, then a volley of unintelligible cursing. This goes on for about a minute and then calms slightly as a coherent sentence constructs:

'Yoo bastard Mandrrrakk. I know it's yoo. I'm going t' fokun' razor ye when I catch ye. Yoo hear me, ya fokun' git'ya? Yoo hear me, Mandrrrakk?'

Harry staggers and falls outside the building laughing uncontrollably. He attempts to close the big green doors but he breaks down and sinks to his knees completely overcome with mirth. He manages to stagger up as Radcliff yells his swansong from the depths of the lift-well, just audible through the slowly closing door:

'I'll batter ye. I'll fokun' razor ye. I'll fokun' bottle ye. I'll–'

Mercifully, the doors slam shut. Harry manages to lock the doors then staggers, still laughing uncontrollably, into the near-deserted car park. Two cars: Harry's silver Rolls Royce and Radcliff's maroon XK Jaguar. Harry collapses across the bonnet of the latter, splaying out the big bunch of keys, still held firmly in his hand, onto the highly waxed surface. When he manages to stand he gouges the larger of these keys along the length of the Jaguar in one continuous movement, etching deeply into the immaculate paintwork and down through the primer to the metal. He finishes the stroke like a discus thrower, hurling the keys out into the pitch-black night. Then he's off in the direction of Heathrow, weeping with joy.

Recovered from his merriment, Harry checks in at the flight-desk. The check-in girl looks up his ticket number on the computer.

'I'm expected in the VIP lounge.'

'Sorry, Mr. Mandrake, your party had to leave a day early. They left a message, they'll meet you at Kennedy airport.'

'No need to worry your pretty blue eyes, Carol,' he reads the name from her badge. 'Just point me in the direction of the VIP lounge bar... I need a stiffy. Can I offer you one? Ha ha.'

'Not just at the moment, Sir.' says Carol, grudging a slight smile, 'If you follow that sign you'll find the lounge bar. Have a nice flight.'

Harry turns and walks off. Carol eyes him with more than a passing interest. He turns back quickly, catching her glance and he tips an imaginary hat, then walks on towards the bar.

A short, wavy-haired man dashes up to the desk and speaks to Carol, feigning shortness of breath. 'Have I missed Henry Mandrake? I tried to catch him before he checked in... I take it he's still on the Kennedy flight?'

'Sorry, Sir, I can't give out that information. Are you

checking in?'

'What?' He glances past her to where Harry's luggage is still on the ramp. He squints his eyes to read the tags.

'I said, are you checking in, Sir?'

'Yeah... I guess. Yes, sorry. Yes.' He hands her his ticket.

She reads it and then pushes it back to him with an air of distaste: 'Sorry, Mr. Casey, you're Club, this is First-class. If you want to upgrade, go to the next desk.'

Harry relaxes snugly inside the majestic Concorde's first class section as it slips effortlessly and noiselessly through the sound barrier. With the terror of take-off forgotten, Harry flirts outrageously with the hostesses. The drinks flow and the company is good. Subsequently, on arriving at Kennedy airport, a mere three hours on and the terror of landing quelled with drink, Harry is, to quote the old nautical adage, 'three sheets to the wind'.

Rex and Hamish stand waiting at arrivals. When they see Harry coming they conclude the obvious.

'Oh my God, he's canned,' says Rex, covering his eyes in exasperation. 'Over here, Hal.'

Harry waves, makes a last farewell to the hostesses then oozes over. 'Boys, how nice of you to meet me. Where's the old barsey-warsey, I've got a fearful thirst... a bit dehydrated... need to lay the dust?'

'Damn it, Hal, I think you've had enough booze, don't you? We've got work to do.'

'Steady on, old love. What did you expect? I mean, don't blame me, the bloody drinks were free for God sake. Had to get the old monies worth... steel myself for the landing. Flimsy looking thing, how the hell it stays up there I'll never understand. Anyway, you paid... waste not want not, as my illustrious uncle used to say.'

'We'll take you to your hotel. We got you a Lincoln saloon, but you're in no fit state to drive.'

'A Lincoln saloon? A Lincoln. What the devil do you take me for, some out-of-town bloody cowpoke? Has it got buffalo horns on the bonnet?'

'What the hell's wrong with a Lincoln for Christ sake? I drive a Lincoln.'

'Well, ex,actly. Look, old love, you don't buy a racehorse and give it a bloody coal-cart to pull, now do you? No, a Roller, I think... Bentley at least... not new... I don't like new things. No hurry, tomorrow will do. Now for heavens sake let's get out of this bloody awful place.'

Harry marches off ahead. Rex and Hamish look at each other in disbelief then follow. When they catch up, Harry's mood has changed.

'Right,' says he with an air of authority, 'if that's sorted we'll have breakfast, then I'll need an hours kip... sleep, that is. Then I'll need two Mackintosh PowerBooks, version OS X, two PCs, any sort, with the latest Windows plus all the old versions 3, 2... mouse and full expansion capability and FuzzyLogic–'

Rex holds up his hand as to say hold on, 'Jesus... anything else?'

'Yes, a JVS 4000 slimline VCR – have you got all that? – Plus a Sony NiCam black screen with voice-activated remote, two Akai DATs with eight Bose AcoustiMass at optimum sound points, and a Krupp's espresso coffee-maker. Oh yes, most important, a couple of girls, IT technicians... pretty, of course. Think you can manage all that? I'll be ready for you by, let's say 2200 hours?'

'Let's say one technician – gender as it comes, and let's say 1400 hours.'

'I don't think so, old luv,' says Harry dismissively. 'I must have the female logic on this, do you see? And we'll split the difference, let's say 1800 hours.'

Rex leans into Harry's face. 'Listen, Mandrake, I told you, don't fuck up. You said you'd be ready – What's the problem?'

'Don't get excited, Rex, I've been ready for years. What I've got to do is put it into, shall we say, layman's terms... It'll take a while.'

Hamish steps in front of Rex and spits his words into Harry's face. 'Why you jumped-up Limey punk... A rust-bucket leaking deadly radioactivity that crashed back to Earth, killing its entire fucking crew and God knows how many civilians. Is that what you mean by 'layman's terms'?'

'We've been to the moon six times,' adds Rex. 'We're light years ahead of your country's crummy technology.'

'I'm not talking technology, Rex,' says Harry, smiling, 'I'm talking Alien life-form. Look, boys, don't get upset–'

'We are not upset, Mandrake,' growls Hamish, 'we're damn, fucking mad.'

'Look, Hamish old luv, swearing is not going to help us here. If I just give you my material it won't mean a thing... that's why I came along. It's only when you've got my conjecture, tagged with evidential fact that it starts to gel. For Christ sake, there are only a half-dozen people on the planet that believe the 1950's Mandrake Experiment is fact. Most of the thinking world sees it as fiction, the stuff of books, films, and television – "the stuff that dreams are made of" – And some of those same 'thinking people,' let me point out, think the same about your... how many was it? six moon landings: fiction. Faked. – I mean to say, look at your faces. I can tell you're having a hard time believing it, and you've bloodywell seen it with your own eyes.'

'We believe it,' says Rex, calming slightly.

'Rex, I've got to convince not only you, but all of them. Now let's go, I'm bloody starving.'

Harry walks off towards the Angus Steak House. Rex gives an eye-rolling glance to Hamish and follows.

CHAPTER SIX

The monolithic 1920's Ridley Building, home to the Carnegie Space Agency headquarters, stands as sentinel of good taste to a chaotic jungle of angular skyscraper appendages. The interior—still in its original art-deco furnishing and fittings—contrasting sharply with the hi-tech equipment and attractive, scholastic people there employed.

Rex enters the main hall, followed, a step or two behind, by Hamish, who is busily attempting to placate a grouching Harry.

'Will you cool it, Hal? You're always complaining about food. Dear God, you had a steak sandwich just two hours ago. Didn't they feed you on the plane?'

'Just catching up with the jet-lag, Hamish, old luv,' says Harry, patting his stomache, 'and I never eat when I'm flying… upsets the old Derby Kelly, do ya see?'

'Oh, drinking a goddam aircraft dry of gin and champagne is okay, is it? And don't call me 'old love'.'

Harry is about to retort. Rex moves between them. 'First meet the team, then we'll have a working lunch, then straight to it.'

Harry's mood changes as he takes stock of his new surroundings. The place is bustling with attractive women. He is introduced to the numerous nubile ladies, all of who appear taken with his genteel Englishness, which, on his realising, Harry exaggerates with cavalier charm. He is now in top flirtatious form, appreciated, seemingly by all, especially a gorgeous black-haired, blue-eyed technician, Kate Ottman, standing a little way back from Rose.

'Rose, this here is Henry Mandrake,' says Rex. 'Mandrake... let me introduce Rose Hawkins.'

Rose smiles, 'Hi, Henry'

'*Harry*, please. I hate Henry... but I insist you call me Hal. I've heard so much about you, Rose.' He takes her hand and kisses it. Rose is impressed and, for a moment, slightly flushed. She feigns comedy to hide her embarrassment.

'Why, thank you kindly, Sire.' She mocks a slight curtsy. 'I guess you've heard a lot about me because there are two of me. You'll be meeting the alien host, Rosette, as these morons call her, after lunch. We thought it better that way... she has a bad affect on some people's appetite. She's in sickbay.'

Harry looks puzzled. 'Sickbay, Rose, how come?'

'She's very sick... cancer... of a sort. An acute cellular disorder.'

'I'm sorry.'

'I'm not,' says Rose, with a flash of malice.

Harry shrugs. Rose looks away... the conversation seems to stall. He takes a small gift-box out of his inside pocket and offers it. 'I've brought you a little gift, Rose. Been in the family a hell of a time, not expensive, not the family jewels, ha ha, but I'd like you to have it.'

Rose is totally overwhelmed – a simple costume-jewellery necklace, seemingly of some great age. As she puts it on she leans to Harry and kisses him on the cheek. There is a bit of chemistry mixing – Rex looks at Hamish and rolls his eyes. The other women also notice, especially Kate Ottman.

Rose smiles and leads Harry, Rex, and Hamish off to

the restaurant where they are to meet Major, leaving the group of women to their tittle-tattle.

'Jees' what a dreamy guy,' says Kate, barely letting them out of earshot, 'I could eat him alive. How about you, Helen?' They all laugh.
'Me?' says Helen, 'Well... for my money, he's a fag. No regular guy talks like that.'
'Or dresses like that,' adds the third woman, 'An' if he's not a fruit, then he's a goddam dyke.'
'Talking of dykes,' chips in the forth woman, 'Rose seems to like him, fruit or otherwise. What you make out of that, Kate, what d'ya think?'
'I'll tell you what I think: I think she and her zombie-dyke sister really do think men are to eat.' She turns to Helen, making a personal jibe. 'Is that right, Helen, is Rose a dyke, you two were buddy-buddies, once?'
Helen looks embarrassed, 'Oh God, don't start that again. We were not 'buddies', we were friends for a while, when I first arrived here – Christ, I'm friends with you, don't mean I wanna slam into your goddam pants.'
'You offering, Honey?' says Kate, laughing and fluttering her eyelashes.
'Gimme a break. We just did some shows and stuff. Leastways, you can find out if he's gay,' says Helen, forcing the subject on from herself, 'you're going to work with him. You gonna try it... Satchel-mouth?'
Kate, taking the retaliatory jibe in good humour, waggles her tongue in her ample mouth, 'All the better to eat him with, Grandma.'
They all laugh.

Joseph Styles, the commander of EarthlabOne, a tall

executive-dressed man, normally of steely and controlled temperament, sits nervously at his roll-top desk in the circular command room. He is staring, lost in troubled thought, into a hand-held mike waiting for the line to connect. The moment it does he rams it to his mouth and gushes, 'Major, we're a man down.' A moment passes. 'Did you hear what I said? We're a man down.'

'Are you certain?' booms Major's voice over the radio-link.

'Yes. He's been missing for at least thirteen hours – he was off duty so it could be as long as twenty-four. We've checked all the suits, they're all present and accounted for.'

'Have you checked the SBS?'

'Yes. We've checked again with video on. He's not moving about, I'm certain... I wouldn't have said if I wasn't. There's something else, Major. We've found decomposed tissue, human residue similar to that in the Junairo... we're still testing it. And something has been jettisoned without authorisation. That's where we found the residue, in the tube.'

'You have to take precautions–'

'Jesus Christ, you don't have to tell me that. The whole crew is crapping themselves... the can-seat hasn't cooled all day. I want a complete change of crew, Immediately.'

'No way.' yells Major, still sitting in the agency restaurant, 'No fucking way. We're coming up – Listen to me Buddy, you're under contract. Shape up or you'll lose the lot. You hear me?'

'A new crew, Major – You hear _me_?'

Major smashes a hand down on the dining table and hisses into the phone. 'Now you get this: No-one gets access to the SBS. Keep everybody off. I don't care what work's being done. Close it down. If you won't do it we'll pull the plug from down here. You understand what I'm saying?'

'Yes. Do you understand what *I'm* saying?'

Major calms slightly. 'Listen, Joe, this affects us all... all we've worked for.' He is silent for a few seconds. 'Okay... Now, get as many people together as you can. Try not to leave anyone alone, keep your people in threes if you can. We must keep up morale. I'll be with you in ten hours. Keep this line open. I'll be in constant communication. That's it for now. Come through for me, Joe... you won't regret it. In ten hours – Out.'

Major puts the receiver down, then picks it up again and makes another call. He speaks immediately it connects, 'Make ready the Orion. And there's a special job I want you to do.' He slams the receiver down again and stares icily into infinity.

Across the crowded restaurant, Rose and party approach. Major shakes away his thoughts and looks up. 'Rose... and this must be Henry Mandyke.' They shake hands.

'Man, drake...' corrects Harry, '... drake.'

'Drake?' says Major, looking at him bewildered. They take their seats at the table. Major continues to all. 'We've got a problem. We got to go back to EarthlabOne. They've lost a man, one Erick Ronan... a weather technician... Anyone know him?' Nobody answers, they all look at each other. Major continues. 'Okay. Now, here's the bit, they've found decomposed flesh residue, same as the Junairo.'

Rose looks shocked. 'Exactly the same?'

'Yes, exactly the same... Sounds fishy to me. The whole crew is on the verge of panic, so the commander says. They all want out. We've got to get up there and show some grit.' He looks from one to the other. 'And understand this, if this gets out we're finished. If we miss a Mars rendezvous, that's a whole goddam season wasted. And after that elapse of time we probably won't be needed.' He studies Harry for a moment. Harry nods understanding. Major continues. 'They say he's been missing thirteen to twenty-four hours, that means he possibly stowed away with us. I reckon he stole some of the sealed containers and scattered residue in the tube to fuck us up, to throw us off the scent. Probably took some copy data. It would be worth a fortune to the international market, to say nothing of the press. I don't think he could have got shots of Rose and Ro–' he stops and corrects, 'the double. You were never together on the station, at least not where Ronan would get access. Were you, Rose?'

'No, I don't think so... maybe.'

Major makes a fist and hammers his forehead. 'God, if he's got a shot of the two of you... Okay, this needs sorting. None of this gets out, not to anybody. One word could finish us. We must contain it.' He looks to Hamish. The big man stands and without word walks off. Rex follows.

Major leans towards Rose, takes her hand and speaks softly. 'Do you mind going back, Rose? I wouldn't ask but the whole goddam sack is about to split... As I say, we got to show that crew some grit. We must go.'

'Suits me – the further away from the freak the better.'

Harry, feeling uncomfortable at Major's intimacy with Rose, takes her other hand. 'What is the latest on the host, Rose?'

Rose pulls away from both hands. 'Very sick, Harry,

she's dying, thank God. You'll be okay for a day or two, Kate will show you around?'

Harry nods. 'We'll take a rain-check on that dinner, yes?'

'What? Oh, yeah... yes,' she smiles, then looks to Major, slightly embarrassed. She shrugs and turns back to Harry. 'All that you asked for, Harry, has been installed in your office. Your 'Roller' as you call it, is in your own park-bay. It's got your name on it – the bay that is. Oh, and you have full accreditation.' She hands him some credit cards and car keys. 'Make the most of the time, Harry... we need you desperately.'

Harry lifts her hand offering the cards and keys, and kisses it yet again, at the same time deftly palming the offered.

Rose smiles, 'Hey. Quit that. You're making Major blush.'

'See you later, Rose. You take care, now.'

Rose and Major start to leave. They stop at the swing doors, to talk to Kate, who has just entered. Rose looks back and nods another farewell to Harry, and then they are gone.

Kate joins Harry, giving him a sensuous, 'Lo, Harry.'

'Kate. How nice. I understand you're going to work with me. I've got a little pressie for you. It's been in the family for ages. Don't worry, it's not the family jewels.'

CHAPTER SEVEN

The *Bronx*, New York City: A filthy, near deserted street, offering an array of drunks, lowlifes and vagrants sitting and lying in doorways. The graffiti covered walls still showing pre-aerosol relic tags: *Aggies, Baldies*, and *Shamrock Bitches,* amongst others.

A scratchy rendering of Hendrix, of *All along the Watch Tower* is just audible, emanating from a disused warehouse. Inside the decaying building the music is now blaring, an ode to a living nightmare, serenading the flotsam of society, who stagger, sing and dance in disunion to the music. Drugs and cheap drinks are for sale behind a makeshift bar, which is overlorded by Nathan, a huge, fastidiously dressed black man. As the congregation dance their hellish quadrille another man, for all-the-world a latter-day Comanche, sans tomahawk bow-and-arrow, bursts in from the fire-escape window and calls to the barman.

'Cops... Bust!'

Nathan grabs up money and packets of dope and follows the second man out of the window. In an instant the place is crawling with cops. Doors rip off disused offices and storage rooms, one after the other. The last door busts open under a shoulder-charge. A swarm of flies momentarily obscures the policeman's view. Gradually the flies settle and he makes out the decomposing remains of a corpse.

'Mary mother of God... mercy on us.' The policeman

recoils in shock and disgust, crosses himself and vomits. Two other cops enter and recoil at the grisly scene. The a third, a woman cop. Enters and kneels down and inspects, using the tip of her pen, to the further revulsion of the other three.

CHAPTER EIGHT

Late evening, edging midnight: In his office, Harry and Kate Ottman are still working. Coffee cups and paper litter the once-palatial Carnegie Agency office suite. Harry is engrossed at his keyboard. Behind him a cine-projector runs the old black-and-white footage of the 1950's *Mandrake Experiment*. After a few moments the projector runs out of film but continues running, flapping the loose redundant end and emitting a monotonous chatter and white strobe-light. The monitor screen Harry is working at is suddenly obscured. Kate, behind him, has removed her blouse and draped it over his PC screen. He stops and studies the garment, compelled to speculate as to the carnal delights exposed behind him. She, completely naked, leans over the back of his chair, cups her huge breasts in her hands, lifts and places them either side of Harry's head, smothering him in her softness.

Kate—now as orator—begins a narrative, a move-for-move blow-for-blow account of Harry's class act performance of eroticism; it is the morning after the night before. Kate, with gusto and lewd hand-gestures re-enacts her night of bliss in Harry's tender experienced hands. She is speaking to her little coven of friends sat round her in the Agency restaurant over coffees, joyfully aware that Rose and Major are sitting within earshot at an adjacent table.

'And then,' continues Kate, giving sideways glances to Rose, 'would you believe, he did it to me all over again.

Darling Harry Mandrake may dress like a fairy and talk like a fruit, but oh my, he flies like an angel.' The women dissolve into laughter. Rose slams down her coffee-cup, gets up and, with much drama, marches out past the giggling group. As she exits, Harry enters. They meet in the doorway. Harry is pleasantly surprised.

'Rose,' he beams, 'When did you get back? I–'

Harry is sent reeling from an almighty smack in the face. The blow spins him around in bewilderment and into the group of sniggering women. This display of unbridled anger only rekindles their suspicion that Rose is carrying a torch for Harry and vice-versa, and a further suspicion that Harry is now carrying the beginnings of a blackening eye.

It is now the day after the morning after the night before. Harry, sporting dark glasses, is relating his Mandrake data in Major's office. Playing on a VCR is the old footage of the British 50's Mars Expedition and the aftermath, now converted to videotape. Rex, Hamish and a starchy Rose are in attendance. The tape ends. Harry momentarily lifts his glasses and attempts eye contact with Rose, to no avail. The meeting ends abruptly. Harry leaves the tape and makes his exit to crushing silence.

Days pass in shunned solitude, different times of day and various stages of bruising to Harry's eye, developing to its peak, red tinged with purple; then on to its gradual decline, a dull brownish ochre. All this time Rose has been cutting him dead in public and physically avoiding him.

Eventually Harry is summoned to Major's office, his black eye now calmed to a slight marbling of maroon and yellowish bruising.

'You wanted to see me, Major?' says mournful Harry

as he enters the office.

Major stands with his back to him, looking out of the window. He speaks at him without turning. 'Your material, Mandyke, was interesting but there are many gaps... there is no footage of the entity or the transformation. This is, as I say, interesting. But frankly it has no conclusion... it is, in a word, incomplete.' Major stresses these last words as he turns and faces Harry. There is now worry as well as dejection on Harry's face as it occurs to him that his gravy-train is about to run off the rails.

'Man... drake,' he corrects again 'Man...*drake*. Look Major, I could give you a lot more. But I'm not even sure if you believe what I've already told you. If I go too quickly it will flow over you and you'll miss the point. This really did happen. My main objective is to convince every one of that. Once I've done that I'll expand into the theory.'

'Let's hope so. Now, I've been checking on a few things. Your science degree you say you've taken, we can't find any record of it.'

'You've been looking in the winning circle, I take it? Wrong place, old sport.' Major gives a troubled look. Harry continues. 'Look, I explained all this to the boys. I didn't get it because I didn't bloody want it. Three times I've sat the damned thing. I could pass it with one eye tied behind my back. The old man's estate provides for me... until I pass out and get married and produce a bloody sprog... only then will I inherit. I intend to do neither. Damn it, I get my fees paid and a bloody good living, and more women students than I can shake a stick at. Why knock it? Your little proposition came as a bit of light relief. So, if there's nothing else...?'

'Okay, point taken,' says Major wiping his hand over his face in exasperation. 'And yes, there's something else. Some of your instructions are a little vague. We appear to have a substantial amount of monies moving through a numbered account in England. What exactly is the product of this, 'William Hill'?'

Harry shrugs, 'You might well ask.'

'I do, well ask.'

'Don't worry, Major, it came to nothing. Let's say it was a hunch. Anyway, I've severed that connection. I have a much more accommodating one now... American. So, if there's nothing else I really must dash. Oh, by-the-by, my contract is sound, I've appointed an adviser, he says it's unbreakable – kosher.' With these wise words Harry, full of rancour, makes ready to leave.

Major moves between him and the door, barring his exit. 'That won't do, Mandyke. We've invested a great deal of money in you. You were warned at the beginning. That warning still stands... more so since the Ronan fiasco. Now, I want you to check out a body that's been found. Two bodies in fact.'

'Bodies.' shrieks Harry in horror, 'I don't know anything about bloody bodies, for Christ sake.'

'I know that, and it would appear you don't know anything about science. The point is you look like you do, that's all you need for this job.'

'But bodies... *dead* bodies?'

'Yes, and believe me they don't come much deader than these. Now, the first body was found a week ago, half decomposed. They put the premature decay down to a combination of the new cheap drug, 'zap' I believe it's called... it's made with the residue scum from refining heroin, all the bums take it, it's as deadly as it is cheap.

And–'

'DRUGS! I don't know anything about bloody drugs.'

Major ignores and continues. 'And... the guy, the body, had aids... so they say. They also speculate he was homosexual.'

'Bloo,dy Nora.'

'Yes, bloody Nora, whoever the hell she is. How they can tell he was gay I shudder to think, but our best play is that aids factor. They found another body this morning, same MO. They want an autopsy. Thing is nobody will touch it, thank God.'

'I don't blame them,' chips Harry, 'I wouldn't.'

'Yes you would, and yes you will. – They've asked our agency to do it.'

'Oh no, not on your life, matey. Not a chance in hell.'

'Listen to me, I want those bodies burned and I don't want any fuss. I don't care how you do it: lie, cheat, or steal, whatever. You're up to all of that, so I'm told.'

'Now steady on, I–'

'Here's your brief and your authority.' Major thrusts an envelope. 'Your neck's on the line, fella – earn your keep... don't let me down. Remember, nothing and no-one is indispensable.'

Harry acknowledges the veiled threat and takes the envelope. Major turns, and without further word offers him the open door. Harry exits, also without comment. The door slams.

Outside, Harry stands for some time in deep thought. He looks down at the envelope then back to the door – to a different door: He now is standing at the top of the steps to Rose's apartment block, studying the names displayed on an old-fashioned handset intercom. He presses the button marked, R. Hawkins, and lifts the receiver. Rose's voice speaks out:

'Hello, who's there?'

Harry answers in his smallest, meekest voice. 'It's Harry, Rose. Please don't hang up.' She hangs up. He rings again. After a few moments, we hear the receiver lift. Rose is there but does not speak. Harry continues. 'Let me talk to you, Rose. Whatever you think of me please listen, I need you for God's sake. The Kate thing was... look, I'm not used to such liberated women. I didn't mean anything to happen... it's you I want. I wanted you from the moment I saw you, and I think you felt the same. Give me a chance, Rose. I'm alone in this bloody place... I need you. Damn it, I need you, Rose. ROSE.' She hangs up again. Harry stands for some time staring into the phone. Eventually, he gently replaces the receiver and walks slowly down the steps. As he gets to the last step the door-mechanism clicks. He turns and dashes back, opens the door and puts his foot in the jamb, then picks up the receiver again, just in time to hear Rose's voice.

'You still there, Creep? Come up.'

This time, Harry tosses the hand set back onto the receiver in his old jaunty style. As it lands he enters the building, at the same time letting out his cry of triumph, 'Ha ha, bingo. Bloody bingo.' The door shuts behind him.

Harry arrives at the apartment door. Rose is waiting for him. 'What can I say, Rose?' offers Harry, sheepishly.

She leads him in and stands with her back against the closed door. 'What can you say?' She leans menacingly into his face, 'How about, *'Ha ha, bingo. Bloody bingo.'*

Harry looks shocked – he is shocked. Rose has just smacked his face, a loud resounding slap. He regains himself and grabs at her and pulls her to him, forcing her onto a kiss. The next moment he is on his knees, kissing the carpet and holding at his groin.

'Christ sake... Rose... what... did you... do that for?'

She looks down at him and imitates his pathetic voice, 'I'm not used to liberated women, Rose' – It's you I want, Rose – Give me a chance, Rose – Ha ha. Bloody bingo. – You moron, you can't even put a telephone receiver down without fucking up.'

'Rose, I–'

'We're not pleased with you, Harry. You've lied to us. You're a shit-house, Hal, of the very first order.'

Harry, still on the floor smarting, looks up. 'I love you Rose and that's the truth. If I'm so bloody transparent you must see that.'

'Yes... I do see that.'

'You do?'

'Yes.'

'And how do you feel about me?'

'We'll discuss that later. Now. I think you want something from me?'

'Rose, really,' says Harry, standing up shakily, 'What the hell do you take me for?'

'Oh. I thought I just made that crystal clear: a rat, liar, cheat, drunk, waster and moron – you want to add anything?'

'Loser? I know, I know... kick a man while he's down, why don't you? I need a breakthrough, Rose. I owe it to the old man, my uncle, for everything he's done for me over the years. I'm not a moron. I've just enough intelligence to realize... that I haven't enough intelligence...' he ponders as to whether what he has just said makes sense or not, '... If you see what I mean?'

'Yes, I see what I think you mean. So, what do you want from me?'

'I've got to check out two bodies, and–'

'I know, and I'm supposed to help you, that's why I let you in. We see the first body tomorrow... it's on ice. Then, I go to see my mother. You're coming with me – I'm not letting you out of my sight.'

'To see the body or to see your mother?'

'Both. We stay at my mother's place for a few days. I have to keep my eye on you. I don't want you running off. Consider yourself under close arrest. Then we see the other body. They've frozen it solid with liquid nitrogen, it'll take them four days to defrost it – 'exactly the same time as it takes from conception to the quickening.' God knows why the jerk told me that, morgue people always say stupid stuff like that, 'a body takes the same time to rot, as gestation,' gallows humour – ironic, isn't it?'

'Bloody hell, that's near on a week. I can't be away that long, I've–'

'You've what? Got something else planned?' she grabs him by the lapels of his jacket and pulls him close, 'Something with Kate?'

He instinctively covers his vitals with his hands expecting another attack. 'No no no. Bloody hell, Rose... I'm just saying–'

'If I catch you with Kate again I'll break your noodle neck. I love you too, you skinny little creep, I think, God help me.' She kisses him, a long passionate kiss, in the middle of which Harry lifts her up and, with great difficulty, carries her into what he hopes is her bedroom. Rose realizing, pulls away from the kiss. 'No. Don't Harry, please. I, I don't. I don't do this. I never... Hal, please...' She is lost in Harry's embrace. They make love in her bed. This is not like before with Kate, this is tender, slow, true lovemaking. For a long while afterwards they lie holding each other. Harry breaks the silence:

'I'm sorry, Rose, I didn't realize I... I was your first. How the hell did you manage to hold on to it all that time?' On realizing how ungallant this remark is he tries to retract, 'Sorry. What I meant was–'

'What do you mean 'all that time?' How old do you think I am for God sake? Christ, Harry, you really know how to sweet-talk a girl.'

'I'm so awfully sorry, Rose. I really didn't mean that. What I meant was–'

Rose's icy stare cuts off his feeble apology attempt. Feeling obliged to give explanation she looks away, slightly embarrassed. 'I just never got around to it, okay? I always wanted my career. Men didn't seem to bother me... I knew I could get any one of them I wanted, so...'

'So beautiful, so, so... modest.'

She punches his arm playfully. 'You know what I mean, creep. Men are easy, they have their brains in their pockets with their change, car-keys, and bubble-gum... and look what I'm stuck with.'

'Yes, what was it? rat, liar, cheat, drunk, waster and moron.'

'And 'loser'... your word, buddy.'

'God's sake, let's call a truce.'

'Okay, truce.'

'I've got something for you, a pressie,' he leans over her, to his jacket, finds a package and hands it to her. She slowly opens it. It's a little gold locket on a silver chain, 'It's been in the family for ages. Keep your pills in it or something.'

'Pills? I don't take pills, I'm as healthy as a horse.'

'No, I mean, *the* pill.'

She looks at him, slightly perplexed. 'Is this how you got into Kate's pants? Don't answer that... Thank you, Hal, I adore it.' She puts it on; it looks beautiful against her naked body.

'By-the-by, Rose, what did you think of my so-called seminar? Major was totally nonplussed.'

'Oh... it was better the first time I heard it.'

Harry looks surprised. 'First time. What do you mean, 'first time'?'

'I heard it before, in Edinburgh. I was there for a month, two years ago. I was intrigued, I went to one of your lectures.'

'You did? You know more about this than I thought.'

'You'd be surprised what I know, and if you'd read your brief like you were supposed to, you lazy bum, you would have known that.' She gives him a look of contempt. 'You're a creep, Henry Mandrake, but you'll do for me.' She slips on his shirt and stands. 'Now, if you're to see some nice dead bodies and my mother, I'm going to have to fix that eye.'

She moves across the room and gets her make-up bag, Harry twists his legs from the bed and sits and watches unable to take his eyes from her. She sits, straddling his lap, and begins to obliterate the slight remaining maroon and yellow bruise. After a few moments, she reacts to a movement.

'Hey. Stop that.'

Harry, smirking. 'What?'

'You know. That.'

Harry lifts both his hands above his head in the pretence of innocence. 'What Rose? What?'

'That. Now, stop it, you'll make me smudge your eye...

Harrrrry. Not again.'

They fall to lovemaking again. Passions spent, Rose's face snuggles on Harry's shoulder.

After a night and half-a-day of carnal bliss, Harry is now noisily throwing-up outside the city mortuary, the first of the two bodies seen and dispatched. Rose looks on with caring eyes.

An hour on, Rose's face snuggles Harry's shoulder once again. Refreshed and the nightmare of body-number-one behind them they are now luxuriating in Harry's Rolls Royce, journeying to her mother's house at Southold, Long Island.

It is early afternoon when they arrive at a little wood-shingled house with rambling unmanaged garden. A gracious middle-aged woman greets them at the stoop.

'Rosey. How nice.'

Harry is cordially introduced to Mabel, Rose's widowed mother, a seventy-year old and obvious beauty in her time. She is soon totally lost to Harry's mild manners and quintessential English good looks. Over dinner, Rose becomes more and more agitated as Harry and Mabel chat and laugh, Harry is getting his feet well and truly under the table.

Later, over the dishes, Mabel confronts her frosty daughter: 'I know you want me to shut up, but it's got to be said, you're not getting no younger, girl. Mark your territory before those high-class bitches you work with get their feline claws into him an' gobble him up.'

'God's sake mother, 'mark my territory,' you make me sound like an alley-cat. Do you have to be so crude? I've only just met him I–'

'You slept with him yet? You slept with anyone yet…

male, female?'

'What the hell are you implying?' says Rose, shocked.

'It don't make that much difference to me, Rosey... all I know is your only beau, of any kind that I know of, was your lady friend at the agency.'

'I had lots of friends at the agency, male and female.'

'Well you need some love in your life or you're going to die a withered old maid. They're going to put 'returned unopened' on you tombstone.'

'Well I'm not... gay, if that's what you are implying, and I've...we've... oh, for god's sake, I don't have to justify myself to you. I'm not a teenager... Jesus.'

'Leastways you're not sleeping with him here if that's what you was hoping. Not under my roof? Make 'em wait... get a ring on your finger, that's my advice, sooner the better in your case.'

'God sakes mother, this is the eighties, not the twenties, he'll hear you.'

'Good, let him hear, daughter. Let him know what to expect, or what not to expect, know what I'm saying?' She wipes her hands on her apron and walks off.

Rose calls after her. 'I've waited thirty-nine years, mother, and I ain't just hoping, so don't bother with the guest-room.'

CHAPTER NINE

A cadaverous moon – like a woman rising from a tomb – moved slowly across the sky. Was she searching for dead things? This night she would be lucky; her silvery fingers penetrating the barred window of the Carnegie Space Agency's isolation ward, illuminating the hardened features of Rosette and a ghastly second figure appendage. It is Cameron, the crewman from the rescue shuttle, squatting on the floor with a look of terror on his face. Between his legs is the body of a male nurse, oozing slime and fluid from every opening in his clothing. Rosette's hands are tenderly caressing Cameron's neck, her fingers penetrating deep into his flesh. The three bodies shudder in a state somewhere between agony and ecstasy.

Rose, in her own ecstasy, nuzzles Harry's shoulder as he drives. She is reliving, in her mind's eye, the past week of carnal bliss, each and every detail. She is oblivious to New York City splaying out in the weak morning sunlight that now condenses with the cold night air, turning it into a light mist and making the city Mortuary, just coming into view, shimmer with an eerie beauty.

They leave the car and walk arm in arm through the ornate iron gates. Inside the archaic building, they present their documents to the attendant, and body number-two is duly wheeled out on a metal gurney. Harry gets a glimpse of the decomposed face as the cot is whisked on into a post-mortem theatre. He and Rose follow. The mortuary attendant stands back and reads off the paper nametag unceremoniously affixed to a lock of the dead man's hair – the body lacking appropriate feet and toes. 'Aldo Fremick,' informs the attendant, 'Got his name from his stuff, it's all sealed in the plastic envelope. I'll get it. You'll find he's just a no-account bum.'

Harry gingerly inspects the grisly remnant cadaver. The bottom half, legs and trunk, are missing. Only the upper torso remains. Harry gags. The attendant returns and hands Harry the envelope. 'If you're gonna throw, please use the bucket.' He directs a bucket in Harry's direction with his foot. Harry takes the envelope and swallows.

The attendant gives a gruesome smile, 'Have you finished, or shall I leave yous for a while longer?'

Harry regains his composure and speaks out as officiously as he can, in spite the dire circumstances, 'Yes yes, finished. You must cremate it immediately. I'll burn his effects after I'm done. You must scrub everything including yourself, immediately.'

'Bet your sweet ass, bub,' the attendant mutters as he wheels the cot to the small adjoining cremation chapel. He parks it alongside a waiting minister with open prayer book and hurriedly leaves.

Rose covers her mouth and whispers, 'We have to see this one burn too, Hal… Major's orders. We must be certain they've both gone.'

'What in hell happened to them, Rose?'

'I don't know. I've read both reports. There's nothing in them to worry us. It must have been the cheap drug, zap. All the bums use it. Dogs or rats ate the rest of him – I need to get outside.'

The body passes to the furnace. Rose and Harry walk through the door leading to a small observation window.

The body burns.

'They've missed something,' says Harry, 'did you notice the lips?'

'Lips… ugh.'

'Yes, the left side of the lips weren't decayed. Everything else, except the lips and gums.'

'You mean you actually got that close? My God, Harry, how could you? I'm getting out of here, I feel sick. Don't mention lips again or I'll throw up.'

Outside the Mortuary – now ugly in the sharp light – Rose gasps for air.

'You okay, Rose?' says Harry.

'I guess. I'm sorry, Hal. What must you think of me? I'm supposed to help you.'

'Take a moment to get your breath.'

'I'm okay now. What next? You going to make the reports, or leave it to me?'

'Not so fast, Rosey. I'll get you a cab, then I'm going to check out a few things.'

'No way. I'm coming with you. And don't call me 'Rosey', I'm no Hepburn and you're certainly no Bogart.'

Harry pulls her to him and speaks with curled lip, mimicking, *African Queen* dialogue, 'Give us a kiss, Rosey, old girl.'

Rose responds, but at the last minute pulls away. She covers his lips and kisses the back of her hand. 'Sorry Hal, I'm off lips at the moment.' They both laugh as they get into the Rolls.

'So, where are we going?'

As Harry drives off he pulls a card from the victim's personal effects envelope. 'The dentist… he had an appointment card in his pocket.'

Rose gives him a puzzled look.

Third world Bronx, mid-day: Harry's car pulls up at a dilapidated building. A group of vagrants laze beside the portico entrance.

A 'Day Clinic' sign hangs lopsided over one of the double doors. Rose and Harry enter under it. Various hand-written signs offer further directions, bringing them at last to the 'Dental Treatment' sign. They enter under it to a passage. At the far end, a man in a white medics tunic looks up from a paper-strewn desk. He speaks one word as Harry and Rose approach.

'Cops?'

Harry nods, 'Yep.'

'What the hell is it this time?'

Harry continues, attempting an American accent that *Dick Van Dyke* would have been ashamed of, 'You gouda patient called Aldo Fremick, I yunderstand?'

'Yeah. Hey, you're a Brit, right?'

'English, old man... on a special mission,' says Harry, somewhat deflated, 'Now, how long ago did you treat him?'

The dentist eyes Harry suspiciously. Rose butts in, 'He's okay.' She gives Harry a pitiful look then continues to the medic. 'He's with us... government.' She shows a badge. 'Just answer the question.'

'Two days ago.'

'You sure?'

'No, I'm not 'sure'... I'm goddam certain. I keep records f' Christ sake. Anything else?'

Rose shakes her head, 'No.'

'Was he homosexual?' adds Harry.

'How in hell do I know? I don't think so, just a bum down on his luck, in pain.'

Rose moves in front of Harry. 'I apologise for my partner.' She gives Harry another censuring look, 'Christ's sake, Harry.'

Harry continues unabashed to the medic, pushing his point, 'Could you hazard a guess, old man?'

'I don't ask. Look, buddy, I don't get paid for this... charity, right. I don't need this. You wanna help, make a donation.'

Rose again tries to apologise. 'As I say, I'm sorry.'

Harry won't be put off. 'Listen, this is very important, people's lives depend on this.'

'Really? Now, if you've got what you want, I'd be obliged if you'd let me get on, people's teeth depend on this,' he taps his watch, 'and as I say, unlike yous two, I don't get paid for this.'

Rose takes hold of Harry's arm, and offers a last apology. 'Thanks for your time.' She pulls the arm and leads him on, protesting. Harry pulls away again.

'Hold on, Rose. I want to ask him something else. Two minutes.'

'On a special mission, old boy – Was he homosexual old boy. Christ. Harry, this is a bloody game to you.'

'No, it's–'

'I feel sick again... the smell. Two minutes... I'll wait outside.' She walks off. Harry returns to the dentist. As he approaches he takes out his money clip.

Rose settles herself in the Rolls Royce and closes her eyes. After a few minutes, Harry joins her. He is carrying a package that he puts, without explanation, into the glove compartment.

'You feel okay?'

'Yes.' says Rose, seemingly still annoyed.

Harry starts the engine and pulls away. He senses her hostility but continues anyway. 'Just two days ago – whatever it was that did that to him missed his lips... it obviously didn't like novocaine. It must have happened just after his visit to the dentist. I've–'

'Stop it. STOP IT,' she shrieks, grabbing at her mouth.

'Bloody hell, Rose, what is it?' Harry he slams on the brakes.

'Stop the car. Quickly, I have to get some air.' As the car stops, Rose opens the passenger door, leans out and gags violently. After a few moments, she pulls back into her seat and slams the door shut.

'You okay?'

'Yes... but for God sake leave it, Hal. If you mention, you-know-what again, I'll throw-up in your precious Roller.'

'Sorry.'

'Look, Harry, anything could have happened. He could have stolen someone else's jacket, picked someone's pocket, it could have been someone else's appointment card... who knows?'

'I said I'm sorry. I–'

'You've done your part. Now just leave it. God, I'm getting sick of this. I should have burned the freak in space when I had the chance. I can't take much more of this, Harry,' she puts her head in her hands, 'I think I'm losing my mind.'

'Okay okay, Rose, okay. I just want to check one last thing, then I'm finished with it. I'll drop you off.'

'No. Not yet. What's the 'one last thing'?'

'He had a club card as well as the dental card, *Nathan Detroit's*... that's John Bunyan, isn't it?'

'Damon Runyon... Jeees, you really are a moron,

Harry.' She looks at him and realises he was joking. She playfully punches his arm and laughs. He zigzags across the empty road as though he's been pushed off course. She laughs louder.

'That's better, Rosey, old girl.'

'Make it quick, Hal. There's something that I need to do, too.'

'Oh. What's that?'

'I've only been doing it for a week or so, Harry,' she gives a seductive smile as she snuggles up, 'I want to make sure it's not all been a fluke.'

'To hell with it, Rosey, your place or mine?'

'To hell with it, Harry, how about spreading a blanket on the back seat. I missed all that as a kid.'

Harry pulls the car over and they climb into the sumptuous back.

Outside Rose's apartment building a mangy tabby cat sits staring mindlessly into space. It scats as Harry's car pulls up. Rose gets out and walks towards the steps leading to the main door. She stops and looks back to Harry, still peering from the window.

'Give it up, Harry. Will you come here tonight?'

'You want to re-phrase that? Ha ha ha.'

'No.' She imitates his lewd laugh, 'Ha ha ha.'

'Wild horses... what?' He blows a kiss and drives off. Rose watches until the car is out of sight.

CHAPTER TEN

The only comparison between Harry's agency-allotted New York apartment and his home in Edinburgh is the collection of ancient and modern IT peripheral, and of course a few mandrake pot-plants. These, scattered around the ultra-modern décor, are his only concession to the antique, plus an ancient roll-top oak desk, at which he now sits working busily making what resembles a shoulder holster out of a leather camera-case.

Finished he takes two large flat-handled dental syringes and a huge bottle labeled *Novocaine* – illicitly acquired from the Bronx dentist – from a drawer. The three-inch long needles rupture, one after, the other through the cork of the bottle filling the two glass barrels, totally ignoring the dosage lines, with the cloudy yellow fluid. He then fits them, suspended by the plungers, into the leather contraption. He puts it on as if it was a pistol shoulder holster, and makes an adjustment, then runs a loop through his belt and adjusts again. He puts on his jacket and turns to face a big gold-framed mirror. His hand deftly grabs in-and-out of his jacket. The hand is now holding one of the syringes at arms length, like a fencing foil. He challenges his reflection:

'I say, old chap, are you damnwell addressing to me?' The reflection snarls. He squirts a small amount of the liquid into his mirror image. The face contorts and snarls back, decidedly nonplussed. It is now the decidedly nonplussed face of Nathan staring back at him, the same man from the vagrants' rave. Hands on hips he stands behind a counter awaiting Harry's order – Harry having just entered *Nathan Detroit's* bar.

Harry winces at the loud jazzy music as he stares back at the huge black man. Nathan gestures to him with a strange pout of his lips and a raised eyebrow that interprets, 'Pray, what's your poison?'

'I'm looking for dis guy called Aldo Fremick,' says Harry, again proffering his New York accent, 'I understand he waza member here?'

Nathan, in spite of his size and villainous look, is, unashamedly homosexual. He looks at Harry in mock amazement, 'Let me see if I understand you, my man – one, you don't want to buy no liquor, and two you don't want to buy no information?'

'Oh... double brandy and keep the change,' says Harry, passing a twenty-dollar bill, 'Now...?'

Nathan pouts again, pours the drink, rings up the till and pockets the change. 'I just love your accent. Bring your drink and follow me.' He ducks nimbly under the bar then nods for Harry to pursue. Harry downs the brandy and follows as Nathan effectively waltzes off in the direction of the washrooms.

Nathan turns, smiling, talking as they walk, 'This way, Sire... past the little boy's room... right to the end. I'll announce you. And if Aldo doesn't do the job, you come right back and see me, Mr.... do we have a name?'

'Harry.'

'How nice... Mr. Harry.' He stops at the door, taps affectedly with his knuckle and sings out, 'Al,do, are you indecent? Harry's here.'

Harry stops and looks over his shoulder. Two men have followed them down the passage.

'Enter, Sire,' says Nathan, pushing it open with an exaggerated swish of his butt, 'Your pleasure awaits,' a shaft of light falls dimly into the passage. Nathan's hand extends in welcoming gesture, 'Voilà.'

Harry's hand finds the opening of his jacket. As he walks past Nathan's outstretched hand he gets a glimpse of steps leading downwards. He stops and tries to turn, but the two men are on him. One of whom is the resurrected Cameron. Harry turns to face the first man, who grabs him in a vice-like bear hug. Nathan pulls the door wide open, allowing Cameron to shove the struggling duo down the stairs. They tumble over and over, ending up in a heap at the bottom with Harry on top, locked in the deadly embrace. Cameron turns and walks off. Nathan remains at the top of the stairs to oversee.

Harry, winded and gasping for breath, is staring into a face inches from his own, which now proceeds to exude fluid from nose, ears, and eyes. The mouth sags open expelling a stream of bloody slime. A woman's high-heeled shoes now step into the widening slime-pool, stopping inches from Harry's nose. He looks up. It is Rosette – there is no mistaking the soulless look in her eyes. The man's bear-hug grip slackens and Harry rolls free. He immediately leaps to his feet, grabs the remaining syringe and holds it at arm's length. The other syringe is sticking out of the putrefying man's chest. Rosette eyes it, acknowledging its destructive quality. In lightening speed, she whips an arm across Harry's body. With his schoolboy fencing skills, he just manages to parry the blow and offer an *inquartata* lunge with the syringe. The needle penetrates, but before he can deliver the full dose Rosette recoils, snapping the needle and sending the jet of novocaine into the air. She grabs at her arm shrieking in agony, then backs off into the shadows, dangling the stricken limb by her side.

Terrified, Harry flees into the near darkness of the cellar stacked full of crates and boxes. Nathan follows close behind. Harry comes to a dead end in a corridor of beer cases. Nathan stands menacingly the other end, blocking Harry's escape. He backs up against the wall and, in panic desperation and hysteria, hammers at the whitewash with the heels of his clenched fists, one of which still holding the broken syringe. To his amazement and joy the flimsy partition wall cracks and splits. Harry leaps through, disregarding life or limb. Nathan follows.

The other side of the wall is an identical storeroom of cans, bottles, and cases. Harry dashes through yelling incoherently at the top of his voice. Another wall looms, he hurls himself shoulder first, in the forlorn hope that it is of similar flimsy construction as the other – it is not. It is solid brick. He rebounds from it like a squash ball, smack into his pursuer, sending Nathan crashing heavily onto his back. Harry rolls clear and is up and running again, the frightful fiend, Nathan, close behind. A slapstick pursuit ensues through the alleys of stacked beer and liquor boxes.

Harry, now a few corridors of boxes and beer cases ahead, stops for a desperately needed breather. He carefully sits on a case of cans, all the time listening. He takes up one of the cold cans of beer and rolls it around his sweating forehead, then, covering the can top with his handkerchief, gingerly opens it to a dull hiss. Nathan, oblivious, continues searching. Harry drinks off half the beer then places the can, slowly and noiselessly, on the floor. After a few moments he slowly starts to stand, taking great care not to disrupt the box he's sitting on. He gradually and most carefully takes a step... straight onto the can, he's just placed, knocking it flying.

'Bugger.'

The chase is on again. Harry passes through the hole he had made earlier, Nathan, two-dozen steps behind, follows. Harry sees an open service lift. In ten steps he makes it inside and hits the control button... Nothing. He hammers hysterically at the button. The doors slowly start to move, then mercifully close.

– A crash.

The doors open again. Nathan's hands have plied them apart as if they were made of paper, his head and shoulders now wedge between, hardly out of breath.

'Harry, Honey... m' main man... going so soon?'

Instinctively, Harry hooks a punch at the head, hardly realizing the syringe is still held firmly in his sweating fist. The half-inch stub needle, the glass body and the residue of the novocaine shatter and rupture into the soft tissue of Nathan's temple. The huge head shudders, eyes bulge and the mouth drools with blood and slime as the doors open and shut again. Harry is gone, vaulting the dissembling Nathan and off into the darkness. Leaving the head wedged into the lift, which now starts its belated journey upwards, detaching putrefying head from putrefying shoulders.

CHAPTER ELEVEN

An antique gold-plated carriage clock pings midnight as Harry enters Rose's darkened apartment. He makes his way towards the bedroom, opens the door quietly and stands, straining his eyes to make out the detail in the darkness. As his eyes accustom to the light he stiffens in terror. Something is moving behind him. A white luminous hand reaches out of the shadows and grabs him sharply under the armpit.

'Arggghh..' Harry lets out a piercing yell.

Rose jumps back and lets out a louder, more piercing yell. 'Arggghhhhhh....'

'Dear God, Rose,' gasps Harry, his terror giving way to relief, quickly followed by anger, 'Never, ever do that again.'

'You stupid chicken-livered creep,' yells Rose, 'What in hell did you yell like that for, you scared the be-Jesus out of me?'

'Sorry... I've had a busy night. What are you doing sitting in the dark, I thought you'd be sound asleep?'

Rose switches the light on. 'I've been waiting up for you. She's out. The freak is out – escaped.' Harry just stares at her. 'Harry. Did you hear what I said? She's escaped.'

'Yes... don't I bloody know it? Lift both your arms up over your head.'

'Why?' She gives him a questioning look and obediently lifts her arms over her head.

Harry studies her arms intently. After a moment he smiles, 'Why? So I can do this,' he cuddles both arms around her and buries his head in her chest.

'God, I love you, Harry,' she whispers into his ear as she hugs him, 'chicken-livered or otherwise. So, what are you going to do about the bitch?'

'I thought I'd sleep on it. Ha ha.'

She punches him hard in the arm. He lifts her up and carries her, switching the light off with his elbow, into the blackness of the bedroom.

The blackness flickers away under a circular neon light. Rose is examining herself in her illuminated bathroom mirror. She inspects her face, then leans close and inspects her eyes, pulling down one eyelid then the other. Her hand now touches at her lips tracing their beautiful outline, top, and bottom – 'Ugh, lips.' She covers her mouth and gags, turns and vomits into the toilet bowl. She flushes, rinses her mouth then looks back into the mirror. Harry is now in the reflection. He puts his hands on her shoulders and they stand staring at each other's image.

'I want you out of this, Rose. I think you've had enough. Go to your mother's for a while… Will they let you go?'

'They can't stop me.'

'No. I guess not.'

'Come with me Harry. They don't need you any more, and you don't need *them*. Everything is money to them… I'm sick of it. I gave my youth for them. I'm not giving them anything else… Please.'

'Rose, Baby, I can't. I owe it to my uncle, to myself and to you. You're right, I am a waster and all the other things you said. I just want to do something for myself. I need to change. I need–'

She puts her hand over his mouth. 'I don't want you to change, Hal. I love you as you are... I love you as a creep. Honest to God, Henry Mandrake, you'll do for me exactly as you are. I want to beat you up every time you cheat on me.'

'Rose, I will never–'

'Don't say it, Hal. I know you'll cheat... you can't help yourself. You're weak and I love you for it. I've never loved, even liked any other man. I never will. Come with me, Harry.'

He moves away. 'I'll drive you down to your mother's, but I must continue. Sorry.'

'Don't bother. I'll take the Greyhound. I'll stay for a spell, maybe even a couple of weeks, months... we need some time apart. I don't know yet, but maybe I won't come back here. I need to sort out my life.'

'I'll come and visit–'

'No. Don't come, not unless you are going to stay. Give it up, Harry.'

Major, wearing his grimmest look of bad intention, sits working at his desk. Sentinels, Rex and Hamish, flank to his left and right. They come to attention at a knock on the door. Major grudgingly responds:

'Come.'

Harry enters. Major speaks without looking up, 'Okay, Mandyke, to business: the woman's out... but not for long.'

'Look, Major,' says Harry, his excuses welling on the tip of his tongue, 'what you have to understand is–'

'So... you got beat up in a gay bar?' says Major, now

looking up, 'Well well. Yes I know – the boys have been on your ass the whole time.'

'Yes but–'

'Shut up and listen. You got caught up in a drugs-dispute, right? – Someone cheating on one of the families, right? One man pumped full of zap, the other executed in some gangland ritual – they smashed a syringe into his head... RIGHT?'

'No. It wasn't like that. What happened was–'

'Shut up. I'm not asking, I'm <u>telling</u> what happened, for the record. None of which is open to conjecture, not to anybody. Do, you, understand? – Speak, damn you, DO YOU UNDERSTAND?' He smashes his hand down on the desk as he screams the words. Harry is speechless. Unable to answer he nods a definite 'yes'. After a long silence Major continues, 'Now, Mandyke, what have you got for me?'

Harry is lost for an answer. He has to think quickly. 'I, I'm working on a theory of my... illustrious uncle's. Yes... I've been looking back... through my old lecture material. It seems he, my uncle was convinced the entity had... had made a vital error. Yes...' He stops and thinks for a few more seconds, desperate to gather his thoughts.

'Error, you say?' prompts Major. 'Go on.'

'Yes, an error. The entity had used the intellects of the two crewmembers to create one. He, my illustrious uncle, theorised on the mechanics of the, the... shall we say... the metamorphosis... Yes, that's it, metamorphosis.'

Rex, who until now has been disinterested, picks Harry up. 'Metamorphosis? – Explain.'

'Well, old man, it works as so: Errm... right... part of the flesh is used as building material, and part is used as fuel to generate heat for a living, shall we say... crucible, as he, my illustrious uncle–'

'Jeeesus H Christ.' explodes Major, 'Will you stop saying, "illustrious fucking uncle." God damn it, get on with it, man.'

'Sorry. Where was I? Oh yes, as my... as *he* put it. I personally think it had copied something that was already degenerating. I think the crew was already suffering from some sort of gamma radiation poisoning. That, or the entity used too much material and caused a conflict within... a form of malignant cancer. It seems to have learned from its mistake.'

'I don't think so, Hal,' Rex chips in, 'Rosette is not responding to treatment. We think she's dying.'

'Damn. I was hoping to–'

'You were hoping to what?' growls Major, turning on him with malice. 'Listen, buddy, this is the best news we've had for months. She's crawled off somewhere to die.'

'Believe me,' says Harry, 'that is not the case.'

'Yes, it, is. And that'll be the end of it. I don't want you making waves; we are a very insular group here, understand? It's simply a matter of timing. Once the contracts are signed you can negotiate whatever crazy project you like, with my blessing, with them, the government. But until then I don't want to hear anything more about novocaine, metamorphoses, entities or any other such-like scaremongering, cockamamie fucking nonsense.'

'But it's still out there. We have to–'

Major puts up a hand, cutting Harry short. 'My people are already working on a shield for the next launch.' He turns away, to Hamish. 'You explain, Son, I can't bear to look at him anymore.'

Hamish shrugs and smiles sardonically, 'It's based on your 'illustrious uncle's' own documented account, Hal, the destruction of the 1950's, shall we say, anomaly.'

'So you haven't been a complete waste of money,' adds Major without turning. 'We need your uncle's material, but I'm not sure if we need you. Shape up or ship out, contract or not. This meeting is over. Leave your report and go.'

Harry makes to offer a final comment, thinks better of it and leaves. As he walks out, a short, wavy-haired man walks in. Harry speaks to him as they pass. 'Morning... Mr. Casey, isn't it?'

The man is shocked and embarrassed. He blurts his reply, 'Oh... hi... Mandrake.'

The door closes. Outside Major's office, Harry picks up the reception phone and makes a call to Rose. The line connects. He whispers into the handset. 'How are you? – Me, I'm okay... I don't know, as soon as possible, Rose. I can't, please try to understand, I just can't.' The line goes dead. Harry shrugs and puts the phone down. He looks back at Major's office door, trying to imagine what Casey is up to. He shrugs again and walks off.

Inside the office, Major offers Casey a seat and drink, the former accepted the latter declined.

'Okay, what have you got, Casey? Make it quick, I'm busy. You said something about–'

'The 1950's Mandrake Experiment... that's what they called it back then. I want in, Major.'

'You, what?' growls Rex, in disbelief.

Casey gives Rex a 'who the fuck are you?' look, then turns back to Major. 'If you don't let me in, Major, I'll print what I already have. I just want in... an exclusive.'

'And if I say no?'

'Please don't say no, Major. You should know I've left instructions that in the event of my, shall we say, untimely demise... all is to be published...' he pauses to a pregnant silence, '... I give my word I won't print anything until you say so. I've worked on this far too long not to be in at the end. I can help you, I've studied Henry Mandrake for years, I know what he knows.'

Major looks intently at Casey. 'You believe that cock-and-bull story of a Limey Mars shot in the fifties? There's no proof.'

'Listen, Major, the Brits put a manned shot around the planet Mars back in the fifties, that, is, a, fact. Believe it. I've seen some of the film, the payload was enormous, the craft was the size of a goddam submarine, for all intents and purposes it *was* a submarine. Trouble was it had a nuclear reactor with solid uranium fuel. The British government got cold feet. They tried to stop it, but Lord Melrose – Henry Mandrake's uncle, who was Air Minister at the time – went ahead, anyways.' He stops and looks around. It seems they don't believe him. He continues, raising his voice. 'Look, it went around the planet and crashed back on Earth, that is a fact.'

'So *you* say,' Rex chips in.

Casey eyeballs Rex. 'Who the fuck, *are* you? I take it you're important, boy... the size of your goddam fucking yap?'

'Don't call me, boy,' says Rex, on the verge of laughter.

Casey turns back to Major. 'It's a fact – of the three crew only one survived. All that was left of the other two men was some decaying matter and a lower torso. None of the doors had opened – Sounding fucking familiar?'

'Again, so *you* say,' Rex repeats.

'Yes, boy, so I fucking say. The lone survivor died later, after a dozen or so horrible civilian deaths... bodies drained of all fluid, just crusty dried husks. Now... I know that here in New York, in the past five months since the Junairo incident, there have been a dozen unaccountable deaths... the symptoms are exactly the same. And I also know they've been hushed up under the cover of the aids scare.'

'Did you see any of the 1950's corpses?'

'No, Major, I didn't,' says Casey, looking agitated. 'Look, when I was in Vietnam we had Collins, the war correspondent – You've heard of him, I take it?'

Major rolls his eyes in tedium. 'Yes... Get on with it.'

'Well, he was with us. We went into a village that had been fried two days before. They'd hit it with everything, napalmed the lot. It was still smouldering.'

'Yes, yes, we've all been to Nam. Get to the point.'

'The fireball was–'

'The point.'

'Okay, okay. Bodies were everywhere, people just lying where they fell, completely turned to ash–'

'The point man,' Major smashes his fist onto the table, 'or I'll throw you out of this office, myself.'

'You'll what?' Yells Casey, leaping to his feet and hurling his words back at major, 'You wanna try that, you smarmy fucking jerk? I'll break your damn back. – Why the fuck is it everyone thinks they can dump on little guys? I'm third dan, ju-jitsu. I'll shove your goddam head up your own ass.'

Rex and Hamish come alert, ready to restrain any rough stuff – Rex still stifling his mirth.

Major, staggered by the little man's aggression, proffers an olive branch, 'Okay okay. Christ's sake, will you kindly get to the point… pretty please.'

Casey hovers for a few moments, the veins standing out on his forehead pumping blood for attack. At length he calms and sits. 'Where was I?'

'You was about to shove my head up my ass.'

'Yeah… sorry, I get carried away.'

'One more outburst like that,' chips Rex, 'and you will get fucking carried away.'

Casey totally ignores the remark and continues. 'So, Collins said these burned bodies reminded him of the Mandrake affair. I asked him what he'd meant. He said he was working for an English newspaper in the fifties, for a year. He was asked to cover a rocket launch. Nobody took it very seriously, but because it was outside the Blue Streak/ Atlas agreement with the USA, and with the ludicrous rumour of atomic fuel, they thought it worth a couple of comic columns – you know, see the Brits blow a hole in the goddam Orkney Islands. Collins missed the actual launch because, as I say, Mandrake jumped the gun. But he did see the aftermath.'

'Why did he 'jump the gun'?'

'God alone knows, Major. The Brit government ordered him to pull the plug, but he went ahead anyways. He had the muscle – being Lord of Melrose an' all – so Collins said. He also said that he got to interview one of the crewmembers' wives, and that she confirmed that her husband had pioneered the first space shot. But Collins said he wasn't able to use the story, nobody would touch it.' Casey stops and looks at Major.

Rex, hanging on his every word, prompts him again. 'Continue, continue.'

'Okay, I was in England in sixty-two. I got an interview with the same woman. Now she told a different story. She said her first husband died in an accident, but that it was nothing to do with rocketry or space exploration. I checked her out: After the accident she'd got an undisclosed compensation figure, and her lifestyle had changed considerably.'

'Changed in what way? ... Get, at the risk of having my head shoved up my ass, to the point.'

'Well, get this Major, she's now married to an Air Marshal no less. I was told in no uncertain terms and in the nicest English accent to 'fuck awff. Drop it or we'll drop you.' And that came from a very great height – From the top.'

'The point.'

'Okay... the point: I do this freelance now, I'm not on any payroll. I tailed Lord Melrose until he disappeared eight years ago. Since then I've studied everything, every angle. And this last year I've taken to tailing Henry Mandrake. I'll never let it go, Major. They're out there, the little green men, and they're ravenously fucking hungry. I just want in... or I split the whole fucking sack. That, Major, is... the point.'

Both men stare across the desk at each other. Casey jumps as Major bangs his hand on the desk. 'You won't print until I tell you?'

'You have my word.'

'I'll have more than that. Okay, you're in. I need someone to confide in – Sometimes I think I'm going out of my mind.'

'Great. You won't regret it, Major.'

'I hope not.'

'So, what about Mandrake?'

'Don't talk to me about that Limey fuck. Do you know he's practically blackmailing me? I pay him a fortune and he still bleeds me blind. He's screwing half the women in the team, and half the men too, probably.'

'Yes, Major, I know the man for a libertine waster.'

'You're telling me? He's got a Mafia-connected bookmaker after him... after me. That jumped-up bastard put up the Agency's name as collateral.'

'So...?'

'So, that's what you can do. Stick your head up *his* ass. I want to know everything he does, says, thinks and fucks. The same goes for that woman technician, Kate Ottman. He's damnwell humping her too. Everything understand?'

'Okay... deal.'

'Now, Mr. Casey, what do you want from me?'

CHAPTER TWELVE

A silver necklace entwines Harry's fingers as he drives the Glen Cove road, making towards Rose's mother's house. He studies himself in the rear-view mirror; he looks older, maybe the starting of a slight greying tinge to his temples, maybe the start of a line or two creasing his forehead. Had the past few months of shunned treatment by the Agency personnel finally got to him? 'Naaaa.' he said aloud. He knew what he needed... Rose.

It is now early evening, the sun is still shining but on the wane. The big sky over the Atlantic appears sanguine red, 'Christ's blood on the cross' as the first Dutch settlers would have said. Harry's Rolls Royce turns into the tree-lined avenue leading to the little wood-shingled house that in the setting sun, glowed like amber. Rose is waiting on the stoop to greet him, sitting in a wooden chair and dressed in a scruffy smock dress.

'Are you staying?' demands she curtly as he starts to get out.

'Hallo Harry,' says he, sarcastically speaking the words Rose should be speaking, 'how are you? Me? I'm fine, and yourself, Rose, and your mother? ... Good, good.'

Rose is unmoved. 'Are you staying, creep? Because if you're not, you can get back in your bloody 'motor' and fu–'

'HARRY.' cries Mabel, cutting Rose off mid-expletive,

'It's so good to see you. It's been so long... I thought she'd lost you. Let me look at you... What the heck you doing standing outside? Come on in, come in come in.'

Harry gives Rose a quick look, as to say 'this is more like a greeting'. They all go into the house. Rose is definitely non-plussed. Once inside, Harry gives Mabel a big affectionate kiss on each cheek. Mabel winks at Rose – Rose scowls back. Mabel smiles and now winks to Harry, 'I'll go make us some coffee. I'll leave you two *hate*-birds to say hello properly.' She gives rose a 'buck your ideas up' look then scuttles off to the kitchen.

As soon as she's gone, Rose continues to Harry. 'Are you staying, Hal? ... If you are, I've got something to tell you.' She gives him a contemptuous look. 'You haven't got a goddam clue, have you?'

Harry rolls his eyes, mimicking deep thought. 'Errrm... No no, don't help me. I can get this. Errrrm... Got it, you are the alien entity, not Rosette?'

She looks at him blankly. 'Yeah... you've got it in one.'

He looks at her in mock-disbelief. 'You are?'

'Yes.' A long pause as they stare into each other's eyes. 'You stupid fucking moron,' yells Rose, 'Is that all you can think of? The only alien entity in me was your sperm – I'm pregnant... In the bloody pudding-club, as you say in your stupid, fag-English language... knocked-up.' She pulls the smock dress tight. She is indeed pregnant.

'Oh... thaaat,' says he, as if obvious, 'I knew that. I just thought you were waiting for the opportune moment to tell me – I wish you wouldn't swear, Rose.'

'You lying rat. You–'

Harry grabs her roughly with one hand by the front of her dress, pulls her to him and kisses her... remembering to cover his vitals with the other hand as an expedient. Rose responds momentarily to the kiss. He takes a chance and puts the other hand around her. She pulls away, turns and stands with her back to him, moodily. —A glittering necklace now entwines her neck. Instinctively her hand comes up and holds it up to her eyes, and she turns. When Harry attempts to speak she puts her hand over his mouth and kisses the back of her hand.

'If you tell me this has been in the family for years I'll stuff it into your bloody nose... and I do mean 'bloody'. Thanks, I love it... creep.'

She walks a little way away, head bent, and speaks as she fingers the necklace. 'Hal, I need you... now. I'm pushing forty, six-months pregnant and I'm scared. I don't know what to do no more. All I know is if you don't come now it's over with us.'

Harry sighs, 'Rose, be reasonable. It's important. If we can't contain this thing the whole world–'

'Fuck the whole world... I've given my youth for the whole world... I need you. Christ, Harry you didn't even graduate, what help are you, anyway?' Harry is hurt by this remark. 'I'm sorry Hal, but it's true, they're more up to it. You've done your bit... they've got their shield. Leave it to the experts.'

'I do wish you wouldn't swear, Rose, it really doesn't suit you. Give me a month. They think they've got a shield. They're going to flood the cabin with an electro-charged plasma. They think that will stop it. The idiots aren't even going to test it. I've got to–'

'Then it's over. I don't need you to have this baby.'

'I've got to stay, Rose. I've got to try to stop it. I promise I'll drop it after another month... we'll go to Edinburgh for the birth... anything you like. Please give me a month. You'd be a great help back there, it's right up your street. We're converting everything: film to binary and videotape, air-displacement, moisture, temperature records, sound records, atmospheric recordings, and heat recordings–'

'You already said, 'temperature records' you Jerk,' says Rose, sarcastically,

Harry gives a shrug and continues unabashed, 'Yes, and temperature records, the lot… lock, stock and bloody barrel.'

'Then you don't need me.'

'I do/ we do, Rose. We're taking averages, converting everything to digital, reconstructing it into a four-dimensional résumé. Then they break it down and do the same again. Sometimes I think they're just bloody twiddling their brains – it beats me. But with you there, Rose…'

'Why me?'

'Because you were the smartest on that ship, that's why. It should have been your ride. You should have been captain… I checked the crew's IQs, and you were by far the smartest.' He gives his most appealing look, 'What do you say, Rosey old girl, come back with me… one month?'

'You really are a bastard.' yells Rose, now quaking with anger, 'You'd let your child and me face that… fucking zoo?'

'There's been more bodies, Rose, lots more. She's out there, feeding. For Christ's sake, the whole worlds in danger… our child included.'

'Don't give me that, you damned hypocrite. You just want to prove you're up to it, for you uncle's sake. Well you're not up to it, Hal – come to that, neither am I.'

'But the whole world, Rose.'

'What about aids? The 'whole world' is in danger from that, but it doesn't stop people making love. What the hell can you do?'

'I can try.'

'They don't want you, Harry, don't you understand? It's getting near the big pay-off. Money's talking. You're a danger to them. Remember what they said at the beginning... the chop?'

'But I may make the difference, Rose.'

'Get out now, while you still can, you idiot. They'll kill you, Hal... you'll force them to it.'

Rose stares at him, hoping for a positive response... Nothing. She sighs and continues. 'I have an appointment at the maternity clinic in the morning, Harry. I'm having the new-fangled ultra-sound scan, they say it can see right inside, tell the sex an' all. Will you come? I take it you are staying the night.'

'Yes, I'll stay the night if I'm invited. And of course, I'll come with you, Rose, to see our son.'

'Daughter. I've had it with men – and you sleep in the guest-room.' Harry gives her a kiss, she turns away from him, grudgingly allowing him to kiss her cheek.

A foetus with beating heart. Harry stares unbelievingly at the screen, an arm, a leg, and a tiny face. The nurse squirts another blob of jelly onto Rose's stomach and offers the probe again. Rose smiles and raises an eyebrow to Harry.

Two days on, Harry sits working in his Agency office along with other technicians. He makes a phone call. As he waits for the call to connect he takes out a black-and-white ultra-sound rendering of their child. 'Girl?' says he with a grin, 'Ha. With tackle like that, I don't think so. Hi, son. Hi, Barney.'

The line connects. It's Rose. They argue. Rose hangs up. He waits an hour then phones again, and again. Each time the same, they argue, Rose hanging up to Harry left staring into the silent receiver – technicians snigger behind his back.

Harry, now alone in the office, calls again. He gushes excitedly into the phone, 'Rose, listen. Don't hang up. I've found something. Nobody had noticed it… they're all so bloody clever, pa. The blank part of the original film, underneath there's some new footage. The old man had laid a new layer of photographic emulsion over it and fogged it. I only found it by accident – he knew I'd find it eventually. It's of the metamorphosis, very rough, I can hardly make it out. I've got the lab chaps digitalising it, putting it into binary, whatever. The buggers hardly speak to me nowadays, so with a bit of luck, they won't bother to check it and see that it's new… I'm an annoyance to them now. The good thing about binary is you only see it when you reconstitute it… reconstruct it, whatever. There was something else. One frame is a bunch of equations, what flesh was lost to what proportion was made. I don't quite understand it yet. Anyway, Rosey, they'll have to believe me now. What say, come back, please?'

She hangs up. He phones again, this time, Mabel answers, refusing, politely, to let him speak to Rose. Dejected, he makes another call. This time, Kate Ottoman's sexy voice answers:

'Lo Harry. Where you been hiding.'

They speak for a short while. Harry finally blows a kiss goodbye. Still holding the phone he asks operator for a long-distance call. The line connects. 'Alfie. My god, it's good to hear a friendly voice–'

Late afternoon sunshine sneaking lazily through the trees finds Rose resting idly on the stoop of her mother's house, still in her dressing gown, still very angry, still very pregnant. The telephone is by her side on the table. She sits mesmerised by it, staring at it, willing it to ring. Exasperated, she picks it up and dials Harry's number, only to slam the receiver down moments before it connects.

Wearily she makes the effort to dress, bids a hasty goodbye and drives off in her mother's truck in the direction of the city. Arriving early evening she makes straight to Harry's hotel apartment block. As she enters the foyer she sees a woman resembling Kate Ottman step into the elevator. Rose makes a dash... the doors starting to shut just as she gets to them. She manages another glance at the back of the woman as the doors close in her face. She is convinced it is Kate. Furiously she turns and makes for the stairs, hurling herself up the two flights, retaining the bloated contents of her stomach with one hand and the handrail with the other. She bursts through the fire doors and into the corridor, just in time to hear an apartment door close. Was it Harry's room? She charges up to Harry's door and is about to slam her hand on the oak panel... she hesitates, puts her ear to it and listens. Her face contorts as she strains to hear. A few moments pass... Nothing. She tries her key... the door is bolted from inside. Now enraged she hammers at it and screams out.

'Open this door, Harry, or I'll bust it down. OPEN IT.'

She batters at the door until eventually it opens to Harry, soaking wet and wearing just his dressing gown.

'Rose. What the devil? What's happened? The baby, are you–'

She grabs him by the neck of his robe and bullies him back into the room, Harry instinctively covering his groin with both hands.

'Where is she?' she screams hysterical, 'I'll kill the bitch, then I'm going to break your noodle neck. Where is the bitch?'

Harry pulls away and backs off, trying to tie his dishevelled, sagging dressing gown. 'The 'bitch?' You mean Rosette?'

'Don't try that you pathetic fucking creep. I saw her. You've got Kate Ottman in here, and when I find her I'm going to kick the cellulite out of her lop-sided ass.' She storms around the apartment searching everywhere, raving as she goes, bathroom, shower, bedrooms, kitchen, and balcony, everywhere. All the time Harry is doing his best to placate.

'Honest, Rose she isn't here... Please don't swear. Calm down... just reason it out. You know me, I wouldn't risk it here... give me some bloody credit.'

'What the fuck do you mean, 'credit'?'

Harry giving a face of reason, 'Look, I could take her to a dozen places... why would I risk it here? Calm down Rose, your condition... please, pleease don't swear.'

She screws her eyes at him, 'You think you're so *fucking* smart, don't you? I bet you have had her in a dozen places – Well, I'm not done yet. I can smell her on you.'

'Don't, be, ridiculous. I've just come out of the shower.'

She pushes past him into the main bedroom again, pulling the bedclothes off the bed, inspecting and sniffing the sheets, then rips them away. Harry shakes his head in disbelief. Unabashed, she starts opening the half dozen doors of the huge wall-to-wall wardrobes, starting at the far end.

'For heaven's sake Rose, nobody hides in wardrobes anymore. God, you look ridiculous.'

Rose crashes back the doors, making her way towards the last three. Harry is leaning on the frame of the last door, he shakes his head and smiles mockingly as the penultimate door crashes open revealing nothing but Harry's clothes, the door falling off its hinges. Undeterred she continues to the last door, Harry moves out of her way. Rose is weeping now, floods of tears, as she rips it, exasperated, open to reveal... Kate, crouching naked in the foetal position, eyes screwed tightly closed, body rigid anticipating the blow she knows is coming. It doesn't come. – Harry, aghast starts to move his hands to protect his groin again. He stands rigid.

Rose turns on him, wiping away her tears. 'And don't think you've gotten away with this, Harry. I know you, and I know that bitch. I'm damn sure I saw her in the lift.'

Harry, realising Rose has not even looked into the last opened door, starts to relax. The change in the tone of her voice already conceding defeat. He raises his hands to her shoulders and attempts to move her away. Rose, assuming her mistake, had opened the last door as mere formality.

Kate, her eyes still screwed up, squats rigid hardly believing her luck, then... WHAM. The door crashes closed in her face, and slips askew off one hinge, Rose's last act of frustration as Harry leads her out of the bedroom and into the main room.

Harry, relieved, confused and grateful, resumes his lofty pretence of hurt pride. 'Rose... Rosey, give me credit. Sit down,' says he, oozing authoritative innocence to the now desperately apologetic, weeping Rose. They undress and make love on the divan – Kate voyeur, spying through the crack in the door.

Passions spent, Harry puts on his dressing gown.

Rose sighs as she starts to get dressed, 'I must go home, Hal, I got to get Ma's truck back. I had to come. I had to see you.'

Harry starts to take off his dressing gown. She stops him, 'No, don't get dressed, Hal, I don't want you to see me out. I'll see you soon.'

'But–'

'Give it up Harry, please.'

'Rose, I...'

'...Can't,' defeated, she finishes his sentence, 'I know. Sorry about the damage, sorry about... The agency will pay, of course.'

'Of course.'

Rose leaves. Harry stands at the door and watches her disappear into the lift. He goes back in, closes the door, locks it and bolts it. When he turns, Kate is standing in the middle of the room, still naked, holding a pillow to her front. She twists the bottom half of her body, offering her bare butt to his view. 'You see any cellulite, Harry? – Who goes on top, you or her?'

'You should know, you were bloody watching us.'

'You'd better be careful she doesn't get *you* pregnant.'

'Christ, Kate, I told you, never to come here, didn't I? ... Didn't I?'

Kate gives a mocking smile. 'What's up lover-boy got nothing left?'

Harry eyes her lustfully. 'Oh yes, regrettably, I've always got something left, God help me. A quickie, then I'm off. I'll see you as planned, tonight at the office. And never, ever, come here again.' He walks over to her and smacks her bare rump.

—From a vantage point in the adjacent apartment directly across the street, through military binoculars, Harry and Kate are being observed: Harry's back-lit silhouette walking out of view and into the bedroom, after a few seconds the light goes out. Inside the adjacent apartment, Casey sighs in frustrated disappointment. He steps away from the window and puts the binoculars away, scribbles a few lines in his notebook, gives a wry, unsatisfied smile, then gathers his belongings and hurries out. He leaves the building and makes for his car. As he drives away, another car, driven by Cameron, ominously follows for a few moments then overtakes and drives on.

Around midnight – witching hour approaching – Harry expectantly enters his unlit office. He calls into the darkness in his soppy, singsong voice, 'Katey, hell,ooo. Where are you, honey? Kate, sweetie I can hear your heavy breathing, and I can smell your perfume.' He strains to listen. 'What's that other noise? What are you doing?'

'Over here, Harry, I'm making myself ready for you… You're early.'

Harry tiptoes over to the direction of the voice. 'You naughty girl, I couldn't wait. Where are you?' As his eyes accustom to the light he makes for the stack of metal filing cabinets, the direction of the voice. One large cabinet door is open, as he is about to poke his head around it Kate's head appears over the top.

'BOO.'

Harry jumps. Kate hangs her blouse over the door revealing her large, naked breasts, then stretches out her hands and pulls Harry towards her.

'Hey, you're all wet. What have–' she stifles his words with a big, open-mouthed kiss. Harry responds eagerly, a long kiss, just the door separating them. When he tries to pull away, she holds him fast. After a few moments, he tries again to pull away and again she holds onto the kiss. Harry's hands push at her shoulders but she won't give way. He can't breathe. He struggles. He is now close to panic as he approaches the limit of his breath. Still Kate holds on. Harry is choking. As he lurches back, his feet lose their grip and he slips backwards. Kate's hold loosens. Harry, now bleeding from the mouth, coughs and splutters. He retches as Kate's long serpentine tongue rips out from deep in his throat. The blood coloured tentacle thrashes at him. In his panic he slides backwards on the slippery floor, his feet kicking upwards catching the open cabinet door and slamming it shut tight, wedging the grotesque flaying tongue fast in its steel jamb. The Kate monster is held tight.

From his vantage point on the floor, between the cabinets, Harry gets a full view of the horror. Kate, perfect down to just below her knees, below this her lower legs are unfinished, forming before his eyes. She is fused in a bubbling mass of slime, tissue, and decomposing flesh. Within this slime-pool lies the half consumed body of a man, a torso lying writhing and shuddering in a frenzy of pain and ecstasy. Just visible through the blood-slime is the shock of wavy hair, the remnant of Casey.

From this unearthly union a new Kate, now perfectly formed, rises, a phoenix from her own, and Casey's, ashes.

For a second Harry does nothing, scarcely believing his eyes. Then he's up and running, but going nowhere, his shoes gliding through the widening pool of jelly oozing freely from the heap of offal that once was his old adversary.

Kate swishes her head, to and fro, trying to free the two-foot tongue, ripping at the flaying ligature. It splits and tears, spilling out blood and slime. She is free. Harry's feet find terra firma through the slime, and he's off. He gains the door, so does the projectile tongue. Kate, however, does not. Despite his terror Harry has the wit and control to grab the handle of the door as he exits, and slams it home behind him, once again wedging the tentacle tongue.

In calculated panic, Harry's hands glide through the casement glass of the fire-cabinet to the old-fashioned double-handed fire-axe. He turns, holding the axe above his head, and watches mesmerised for a few petrified moments Kate's jumbled silhouette thrashing behind the reed-fluted glass door panel – the tip of her tongue absurdly dangling Harry's side.

With one fluid movement, the axe passes through the glass panel and lands, up to the stock, into the crown of Kate's head. An ear-splitting, high-pitched scream as the Kate-thing dies instantly and soundlessly... it is Harry screaming.

All is now silent. Harry stands like a pillar of salt, looking in through the shattered door at Kate's body. Her tongue, ripped out from the root with the force of the blow, still held fast in the door jam. She lies naked, her body adorned with jewel-like shards of glass, plus axe appendage. Harry remains inert for some moments. Now something unseen lifts him from his stupor. His face reacts to the rising, revolting, all-consuming odour. On realizing the origin of this pungent stench he slowly waddles off to the lift, heading for the gymnasium a floor below, desperately trying not to displace the unwelcome package swaying pendulously between his legs, suspended, cradled in his underpants.

Ten-minutes later Harry exits the lift, out into reception. He is now wearing an ill-fitting tracksuit, headband, and trainers, and carrying a paper package under his arm. He walks up to the desk.

The lone security guard gives him a quizzical look. 'Mr. Mandrake, where in hell did you come from, and what the heck's that goddam smell?'

'Sweat,' answers Harry with difficulty, his throat still sore, 'Honest sweat, old luv... the smell of the gym... been working out... only time I get is at night, d'you see?'

'What's a matter with your throat?'

'Neck-presses. Now, pay attention. This is of the utmost importance: Phone Major and tell him meet me here in exactly one half-hour. He must come. No matter what argument he puts, he must come.'

The guard bulks at Harry's assumed authority, 'Now you wait a goddam–'

'Just listen,' snaps Harry, oozing authority. 'Have Major wait here for me, in reception. Don't let him or anyone else pass. Do, you, understand? This is of national importance... I have the authority. Nobody passes, right?'

'Yeah, but–'

'I have the authority. Your job is on the line here. Nobody passes, there's a good chap. Now I'm going for a jog. Here in half-an-hour.'

The guard's annoyance at being put-upon is heightened by Harry's patronising manner. He concedes grudgingly,

'Got it, Mr. Mandrake, I got it,' adding to himself as Harry departs, 'you crazy Limey fuckn' jerk.'

Outside the Carnegie agency, Harry starts his jog. Across the street in a waiting car, Cameron watches. Next to him is stone-faced Rosette. She rolls her eyes in dire disappointment as Harry exits. Cameron slams the steering wheel in frustration and then drives off into the night. A short way from the building Harry dumps his unwanted package into a big metal trash bin.

Twenty-five minutes later, Major bursts through the great doors, Rex and Hamish close behind. Major yells at the guard, 'Where the hell is he? I'll kill him with my bare hands. I'll–'

Harry, still in the tracksuit, enters. 'Major.' he calls, 'Good, you're on time.'

'Mandyke, you fuckn' imbecile. What in hell do you mean by dragging me out here in the middle of the night? It had better be worth it man or I'll have you certified.'

Harry, still racing blood from his ordeal with Kate, explodes at Major's unexpected bad temper. 'DRAKE. for Christ sake. Man,drake. Get it right, you blasted moron. It's been eight bloody months, damn it, DRAAAKE.'

'Drake'.' says Major, puzzled, 'What the hell are you raving about, man?'

Harry closes his eyes in exasperation, 'Follow me, I have something to show you.'

Major, Rex and Hamish follow Harry into the lift without further word. In the confines of the elevator Hamish, sniffing the air, gives a quick suspicious look to the sole of his shoe. The journey up is most uncomfortable, the four trying to avoid looking at each other, three of which pondering the origin of the offending smell, the other pondering the futile explanation he is about to offer.

Harry, now dressed in tailored suit, shirt and tie walks immaculately and warily up to Major's office. He makes three abortive attempts to knock, at the forth, claiming a morsel of valour, he taps gently, it is the morning after the night before.

'IN,' booms Major.

Harry enters. Major is sitting stoically behind his desk, Rex and Hamish standing either side. Major speaks without looking from his papers.

'Don't sit down – this won't take long. We've tidied up after you.'

'Just hear–'

'Don't speak,' yells Major across the office, 'Don't sit. I don't even want to look at you. Just listen.'

'I cleaned up your mess,' adds Rex, with malice.

Harry again tries to speak, 'Just hear me out–'

'Will you shut up,' hurls Major. 'None of this gets out, understand? Now, we've got a month to go of the present contract. The Russians have started their own Mars programme so we may just be able to put this mess down to industrial espionage, maybe. It's up to you.'

Harry gives an angry, puzzled look. 'You can't possibly think of brushing this under the carpet, Major,' says he in disbelief. 'Kate was supposed to be a plant. We'd never have known if I hadn't got to her before she finished metamorph–'

'Jeezus. Will you shut up.'

'But it's out there bloody feeding for God's sake.'

'You have nothing on for the next few days,' says Major, ignoring, 'no more fucking murders planned?'

'What. Why? I don't like the bloody sound of this.'

'We have one last task for you,' says Hamish trying to hide his look of contempt, 'We want you to check out another body.'

Harry is shocked to silence. He takes two quick steps towards Major. Hamish leaps the desk and grabs him in a bear hug before he finishes a third.

'No way,' screams Harry through restricted larynx, 'I won't do it... I won't. Not a hope in hell.' He struggles, kicks, and weeps.

Major unmoved. 'That's just about what you've got, buddy, a hope in hell. You'll do it because you've got no other choice. The boys have all your notes.' He leans forward and places a bundle of bills, papers and receipts onto the table.

Harry shrinks at the sight of them. 'I can explain these. They–'

'No, I can explain these. You thought we were easy pickings. You're in big trouble, buddy.'

'But–'

'A Mafia-connected bookmaker – you owe them money. Maybe I'll let them have you, save us the trouble.'

'No. You can't–'

'And you're in trouble with us. You sold us out to that lunatic, Casey. And you're in trouble with Narcotics. You corrupted and bought drugs and medical apparatus illegally, from a state registered dental surgeon.' Major stares. Harry has to look away.

'But best of all, Hal,' says Rex, now taking up the inquisition, 'we've also got you on a murder-one rap. The police have got Kate's body on ice and we've got the axe with your prints on it. They'll fucking crucify you. You ripped her tongue out while she was still alive then you killed her with a single axe-blow to the head. You'll do it.'

Tears of fear well in Harry's eyes… moments pass and he is openly weeping.

CHAPTER THIRTEEN

The interstate jet was small, grubby and looked well used, which to Harry, with his dislike of new things, was marginally reassuring. He sat quietly in his seat in deep thought. This time, there was no flirting. He took his huge brandy without a second look at the pretty stewardess who served it, and drank it off without tasting it. His eyes carried the same tears and fears they had when he left Major's office six hours previous, plus, as he sat alone in the crowded aircraft, a look of utter dejection.

Harry now drank dark, brackish coffee from a chipped enamelled mug. His sad face had not changed in spite of the change of location: New Orleans airport police precinct. He sat forlorn in a small heavy wire holding-cage situated dead centre of a large deserted room – a room within a room. After long hours of utter silence Rex and Hamish enter, accompanied by a huge gesticulating police officer.

'God-damnedest thing I ever did see,' says the officer, in a high-pitch twang, 'That there good ol' boy,' he points at Harry with a handful of keys, 'was riding an aircraft with two syringes stashed in a leather shoulder holster, just like you'd carry a piece… both full of pure novocaine. Shit, the size of them things, took near on half a pint in each. What in hell was he about?'

Hamish condescends to answer, trying his best to confound. 'We'll vouchsafe him… we'll take him back with us.'

'I don't know about that there 'vouchsafe' an' all – we

got a major narcotics violation here. What is this guy to you anyways, he speaks like some kind of fruit foreigner?'

Hamish, feeling he has explained sufficiently, ignores the officer.

Rex, tongue in cheek, answers, 'He's Australian. You've heard of the Flying Doctor? Well, he's the Flying Dentist.' He turns away to hide his stifled chuckle. Hamish gives him a disparaging look.

'What'd hell you take me for, some kind of Hillbilly cretin? Hell with it, take him. You boys' as crazy as he is. Just sign for him an' his hardware, that's all I care about.'

'Okay,' says Hamish, snatching the keys from the officer's hand. The officer looks like he might retaliate.

Rex steps between them. 'It's okay, my partner is a little edgy. We'd like to talk to the prisoner, while you get the papers.'

'Suit yourselves.' The officer walks off.

Hamish opens the door into the cell. Harry, staring blankly at the floor, speaks as they enter.

'You set me up. Why?'

'You pissed in the tent, Hal,' says Hamish. 'You leaked to the press.'

'I didn't put Casey on to you. That little man has followed me, on and off, for over a year... You all knew that.'

'That's irrelevant now.'

Harry turns from Hamish. 'Rex, old man, surely you believe me?'

'We can't let you back, Hal. Seems we got two options: One, we give you to the cops. Or two–'

'Kill me. That's it isn't it? Rose was right. That's why you wanted me down in this Southland bloody swampland. Well, it won't sodding do.'

'Kill you? Hey. That's a third option. We hadn't considered killing you... thanks for that, Harry. And New Orleans is not a 'swampland'. It's the birthland of the Blues. Don't you listen to the songs? He starts to sing Sinatra style, *'They nursed it, rehearsed it and gave out the newsss... that the Southland gave birth to the bluesss.* —You got the blues, Hal?'

'You bastard,' growls Harry, and turns again to Rex, 'I thought you at least were my friend.' Rex shrugs, as to say 'things have changed'.

Hamish leans into Harry's face. 'You said 'it won't do,' Hal. Why not, what did you mean?'

Harry stands up and hurls his words back at Hamish. 'You think you've got me by the short-and-curlies. Well, I have news for you, I have insurance.'

' Insurance?'

'You forget who you're bloody dealing with, Hamish old son. I told you before, I'm not some out-of-town bloody cowpoke. The British Government know that I'm here, and I've kept them up to date. If you look at my fax file – you've got my papers – you'll see a rather large amount of international traffic, to and from my old university. And you know about our universities. Check it if you don't believe me. If anything happens to me, you'll have the Special Branch all over you... and the British newspapers. So you tell Major to bloodywell think again.'

Rex puts his hand on Harry's shoulder. 'Killing you was never an option, Hal. The second option was to keep you down here for a few weeks until the new contracts are signed and sealed. But in light of what you've just told us...'

Harry is terrified he has played his trump card too early? He manages to control himself, realising he is now bargaining for his life. 'Look guys... ha, ha... let's not overreact. It's in a holding situation, I can call it in if I get assurances.'

'Holding situation?' says Rex, looking troubled.

'Yes, holding. Look, I'll go back to Edinburgh, you'll never hear or see me again, I promise. You know me, I've got no axe to grind–' he chokes on the word, 'axe'.

Rex looks to Hamish.

Hamish shrugs, 'It's up to you, partner. For my money I'd finalize it here.'

'Okay okay... it won't come to that. You go back and pass this by Major. I'll stay here with Harry.' He turns to Harry and shrugs, 'I hope we can work it out, Hal. I'd hate anything to happen to you.'

Harry starts to think about feeling maybe a little bit better. 'I should damn-well hope so,' says he, beginning to adjust his clothes and smarten his hair, anticipating deliverance.

The police officer returns with a clipboard of papers. Hamish opens the cage door and Harry sheepishly steps out.

In the late evening dusk, Harry, under Rex's guidance enters the run-down, but still elegant, New Orleans hotel. Rex makes the booking, takes the key and leads Harry up the ornate, dilapidated staircase to the top suite.

'I say Rex old sport,' says Harry with relief as he enters the antiquated apartment, 'this is more like it,'

Rex shrugs, 'You like this dump?'

'Oh yes, it smacks of... decadence. I hate new things. It'll do very nicely.'

Harry, his old self again, takes a drink from the well-stocked mini-bar and offers one to Rex.

Rex declines, 'No thanks. Now, we'll be here for a few days at least, Hal, so is there anything you need? – Oh, the cop gave you these back,' he tosses the leather holster containing the two syringes.

Harry checks that they are both still full. 'Thanks, old man.' He puts them away. 'Are we going out? I take it we have expenses?'

'Yes, we have expenses. And no, we are not going out.'

'Then I will need a few things. Look here Rex old luv...' says Harry putting his arm around the big man's shoulders... Rex squirms uncomfortably.

'Don't do that, and don't call me 'old love' okay. God knows what the hotelier thinks.'

'Ex,actly, exactly my point, Rex,' says Harry, removing his arm. 'If I'm to spend the night with you... you know,' he winks, 'it looks odd... all chaps together. Do you see?'

'What the hell are you driving at?'

'What I'm driving at is... it looks a bit too cosy-wosy, if you take my meaning. What I suggest is... you have a word with the porter chappy and get us some, you know... company. 'Expenses paid' you said... some female chaperones, so to speak. What do you say?'

Rex is relieved. He'd imagined Harry was about to propose something quite different, but he is also shocked at Harry's rakishness. 'My God, Hal, don't you ever let up? What about Rose?'

Harry looks quizzically at his minder, 'Steady on old mate... just a bit of fun. If you can rustle up Rose, fair do's, but she's a thousand bloody miles away. I mean, who's to know? Live a little, Rex, you're a long time dead–' Harry again chokes on his ill-chosen words, the reasons for his incarceration come flooding back to him.

Rex acknowledges Harry's poignant choice of metaphor, he sees the fear return to his eyes. 'Don't worry Hal, we'll work it out.'

'What I meant was–'

'I know what you meant. Okay, you want some company, I'll decline if you don't mind, but I'll dine with you. I like you, Hal. I told you that when we first met in Edinburgh. Nothing is going to happen to you while I'm around, you have my word. Now, I'll order us some dinner. —And for your dessert, Sir, blonde, brunette, white, black, Oriental, female/ *male*?'

Harry balks at the last offer. 'The devil do you mean?'

'Joke, Harry... I make jokes too.'

After his obligatory half-cooked steak meal, Rex retires to his room leaving Harry to prepare for his visitor, his bespoke beautiful call girl. She arrives with a rap-a-tap-tap at the door. Harry opens it with a glass of champagne ready in hand. Smiles exchange. He daintily kisses the back of her hand depositing the glass of Moet into the same and leads her into the room. The poor girl is overwhelmed – this is not her usual salutation. They talk, laugh, dance then inevitably engage in exotic lovemaking. Harry's tender commitment to the sensual multi-climax is enjoyed to the full by both, plus encore after encore.

Finally satisfied in body, and reasonably satisfied in mind, Harry has a last glass of champagne from the cooler by the side of the bed, then yields to troubled sleep... his exhausted companion way ahead of him, lost in well-earned repose. The enormous bedroom is now in darkness, the rest of the apartment bathed in the gloomy moonlight. Rex, wearing just a dressing gown, is sitting in an armchair pulled to the centre of the main room. The rest of his clothing hung about him, tidily stacked on the backs of chairs ready for immediate action. The big man hovers somewhere betwixt sleep and consciousness, eyes half-open and transfixed on the door. Nestling lovingly in his lap – finger on trigger – is a fully automatic Glock machine pistol. The main window is open and the slight breeze sucking at the grimy filigree curtains, flaps them lazily out to the balcony and into the warm night air.

Outside all is quiet, save for the gradual rising noise of the approaching woman's footstep. In the street below the fast walking high-heels strike the stone sidewalk in a hollow, click-clack, click-clack, click-clack, ever closing towards the hotel. The owner of the patent leather stilettos stretches out an arm and pushes the revolving doors, enters and walks up to the deserted reception. Her hand reaches to the register. She turns it to her view and reads the last entries.

A male receptionist comes to the desk from the annexe. He gives a disparaging look. 'Can I help you Madame?' says he, officiously turning the register back to himself.

Rosette, dressed to kill, gives an acrid stare. 'I'm expected,' she hisses as she taps the register at Harry's name.

'Jeeezus,' quips the receptionist, 'Not another one?' She turns without further comment and walks off toward the elevator – one arm dangling limply.

Harry stirs. He awakens his companion with a hidden beneath the covers caress. She responds and they make love again, after which she gratefully returns to sleep. Harry ponders for a few moments then rises, puts on his dressing gown and heads for the bathroom. He enters and closes the door, strips and showers, gargling William

Blake's, *Jerusalem,* through steaming cascading water:
'Anrrrd drrrid thorrrrse feetarrr in anrrrcient tirrrmes
Walkrrr uporrrn Errrnglrrand's mounrrr,tains grrrreen
Anrrrd warrrs the hoooly larrrmb orrf Godrrr
Orrrrn Errrnglrrand's plearrrsarrrnt pasrrrtures srrreen?'

The main room is now in virtual darkness as wispy cloud momentarily hides the moon. Rex is still sitting in his chair. A movement, a vague shape flits momentarily across the room, then it is gone. Rex's eyes are now wide open but showing only the whites, the irises driven upwards into the sockets in agony. His body is fused to the chair with glutinous melting flesh, the Glock pointing aimlessly, held in blackened stick-like fingers, the index baked grotesquely onto the trigger guard. The clouds move on and a shaft of silver moonlight falls into the room. It illuminates in full horror the tortured, corrupting remains of Rex, shuddering, held in lingering death-throws. Harry, oblivious, is now standing in silhouette in the doorway, a towel rakishly wrapped around his waist and a champagne glass in his hand.

He walks into the room, up to Rex and smiles, 'Come to buy you a drink old sport... sitting out here in the dark all on your–' the glass slips from his hand. Without moving a muscle his eyes traverse the room, suddenly darting to the direction of the slight rustling emission from his bedroom. Following this direction, through the door, into the room, to the bed, and on towards the call girl. She is lying in pink contrast to the blue moonlight, asleep, beautiful and dying. A slither of slimy flesh trails across the silken sheets drooling blood. Catalyst Rosette is hunched over her, fused in a delirium of ecstasy. The girl is now awash with the glistening mucus, ears, eyes, nostrils, and mouth, every orifice exuding blood-tinged slime. A movement, a rustle of sheets as Rosette slowly pulls the body off the bed and towards the door to embrace Rex, in the next room, destined to turn one last trick.

Harry, still standing over Rex, is frozen in terror. His whole being fighting the urge to weep and wail as the Rosette/call girl combo slither ever nearer. Somehow he manages to contain the approaching panic. In one chaotic movement he gains the door with as many items of clothing and baggage in one hand, under arm, between teeth, and under chin. His other hand is dedicated to snatching up his holster.

He stands naked in the doorway, his towel lost in the melee. After a parting farewell to his stricken lady friend, and a last look to the grotesque shuddering hulk that once was Rex, Harry empties half of one syringe, *coup-de-grace,* into the big man's temple. Then delivers the other half into ecstasy-engrossed oblivious Rosette and appended hooker, whose groping tentacles now encircle Rex's arms and legs. Without awaiting the grizzly consequence he is off into the hall, half running and at the same time trying to dress in a hop, skip, trip and wobble. As he negotiates Rex's oversized trousers he desperately fights to keep anal-retentive, puckering his lips in tight unison. His feet barely touching the plush carpet as he runs, hitting every fire alarm he passes with hand and elbow, and screaming at the top of his voice, 'Fire. Help... smoke! Fire, FIRE, FIRE.'

Alarms clang and people pour out into the hallway – Harry's warning cry being picked up and added to:

'Fire.'

'What fire? Where?'

'The floor above, I think. I smell smoke.'
'Smoke. What did I tell ya, I said I could smell something?'
'FIRE.'
'Don' panic, DON' PANIC. FIRE!'

Among the panicking hotel guests Harry, half naked and carrying his clothes and baggage, is not at all conspicuous as he exits the revolving doors leaving utter pell-mell behind him. He steps unashamedly, ridiculously, fully clothed in Rex's huge suit, shirt, hat, and shoes, out into the street like a male version of *Annie Hall*. A short way from the hotel he breaks into a trot, looking over his shoulder every second or so. Now running for dear life into the early morning sunrise like the proverbial bat out of hell.

An hour later Harry is sat hunched in an aircraft seat, drinking off a huge brandy, staring blankly at the sky floating past, muttering and shaking with fear.

CHAPTER FOURTEEN

Five hours on, New York is in darkness. In his Agency office Harry is still shaking with fear. He is sitting with a telephone rammed to his ear, the number is ringing but as yet not connected. It has been ringing for some time and he has remained motionless not daring to move, mesmerised by the continuous monotonous peal. Lost in semi-delirium his face is illuminated by a strange, eerie halo of green light.

Finally the phone connects. 'Rose.' He gasps, shaken from his stupor, 'Thank God it's you. Where have you been, your phone's been unobtainable for ages? I think it's being tapped. Be careful – Rex is dead. You were right, they did want to kill me, bastards. To hell with the lot of them, Rose... I've had it... I'm out of here. We–' His incoherent ranting is momentarily interrupted by Rose's voice on the other end urging him to calm down. He calms slightly and continues, gushing, 'I'm sorry, I'm sorry... I'm okay now. She killed Rex, and I've bloodywell killed *her*. I've booked a flight for us to Edinburgh... in four hours... say you'll come. I know it's short notice, but please say you'll come... I need you to help me ... I love you Rose. You won't regret it, I promise–'

'Okay, okay.' Rose's voice yells over his rambling.

'You will? I promise there'll never be anyone else. I... okay, I'll shut up... Right... Right.... Okay. I'll meet you in four hours, at Kennedy. I love... Okay.' He puts the phone back on the receiver next to a bank of telex machines, through which he is running a continuous pile of documents. The green running lights of the machines being the only illumination, hence the eerie halo to his face.

He checks the machines and loads more documents. Now reasonably calmed he opens Rex's attaché-case, the one he'd grabbed from the hotel. As he takes out his customised holster a huge silver pistol drops onto the floor. He picks it up and sticks it awkwardly into his belt. Then takes up the empty syringe he'd emptied into Rex and Rosette, and fills it from a bottle taken from his drawer. He starts to load it back into the holster. He stops and studies it. The second one, the unused one still in the holster, has a clear transparent fluid. The one he has just filled has the normal, cloudy novocaine. He takes the second syringe and squirts out the clear fluid onto the floor and refills it from the bottle. Suddenly, it dawns on him. He has a look of horror in his eyes.

Rex smiles, the big man is just visible in the green light, he is standing half way in the open doorway. His automatic weapon is hanging nonchalantly in his hand. He looks as ridiculous in Harry's ill-fitting suit as Harry does in his – the fly and waist of his trousers stretched open, only the belt keeping his alien modesty intact. 'Hi, Harry,' says he smiling. 'You didn't think that cop would give you back pure novocaine, did you? God, you worry me sometimes. What are we going to do with you? Just let them fall to the floor.'

Harry is stricken rigid with terror. Rex gestures with a little wave of the Glock. Harry obeys and tosses one of the syringes to the floor. It sticks in point first.

Rex smiles. 'Clever. Now the other one.'

Harry tosses the other… but not quite as before. As

Rex's eyes follow the expected trajectory, a little twist of the wrist, pub dart-championship fashion, has the heavy projectile set onto a new target.

BALLSEYE!

The razor-sharp tip finds the soft, dollar-sized area of flesh between Rex's collarbones, just below the Adam's-apple. The weight of the metal grip plunger plus the force of inertia injects half the contents deep into the big man's windpipe.

Rex gasps, chokes and coughs out a huge gob of blood, slime and burning novocaine into the direction of a fleeing Harry, followed by a hale of automatic fire. The random impacts sending sparks and debris crashing around the room. The serpentine tongue leaps from Rex's mouth, slit down the centre by the razor-sharp needle deep in his throat, and throwing curtains of blood-slime droplets into the dust as it thrashes in agony. In his death-throes Rex manages, instinctively, to fit a new clip, which he empties as he dies on his knees, gurgling blood and mucus. The last half of the magazine firing aimlessly into the ceiling showering the room with shards of plaster, white dust and bullet fragments.

Rex is still. His gun raised to the heavens, frozen like some sculptured war monument. Absolute silence, absolute stillness, only a waft of gun smoke lingering around the petrified, slime-caked body. No other movement, no Harry, just settling dust. Total serenity.

A sudden pistol shot. Followed by five more in quick succession then silence again – still no Harry. Rex is held in his petrified pose, his gun pointing upwards as if in some divine state of grace. Gradually there is a slight shift, very slowly Rex's head starts to turn. It stops, leans to one side then lazily slides down his neck and onto his chest. It rests for a second then tips sideways, allowing the syringe to fall to the floor, revealing a line of six bullet entry-wounds. The head slithers off the remnant neck. Harry's six bullets completely severing it from the body, save for a strip of loose skin, causing it to dangle like a swaying pendulum. The fine balance now disturbed, Rex's body pitches slowly forward and falls to one side with a thud, sending the dust up again – Still no Harry.

The middle door of the row of steel cabinets, riddled with bullet ricochet indentations, moves. It squeaks as it slowly opens to reveal Harry, crouched, trembling and cowering. His hand emerges, still pointing the great silver pistol, gingerly followed by the rest of him. His face is a mask of terror.

Gradually this look of terror changes to one of wonderment, and then to disgust as he reacts to a repugnant smell. He looks in disbelief to his lower body. He rolls his eyes as to say, 'Oh no, pleeease, not again.' He slowly waddles off to the lift, to the gymnasium below.

A half-hour later Harry enters his apartment, once more sporting purloined tracksuit and trainers. He quickly disrobes and showers. Once dressed, he places new supplies of documents into the bank of telex machines and starts sending, then he picks up the phone and attempts to call Rose. He sits impatiently waiting for the connection – it won't connect. 'Stupid Yankee-bloody-Doodle, lousy, bastard, rotten phones,' yells he angrily into the dead handset, 'Wake your sodding selves up. Christ, you can put a man on the moon but you can't make a simple, sodding telephone connection.' He rams the receiver back, rips it from the wall and hurls it across the room in tantrum. 'Bugger the lot of you.'

To curb his approaching panic he bites hard on his knuckle. Having calmed slightly he takes up the pistol, reloads it from Rex's case, and sticks it awkwardly into his belt. He then packs a few things and makes for the door, stopping only to give the upended telephone a hard kick sending it to the other side of the room, then walks out, slams the door and heads off into the night.

As he exits the Agency building the first glorious rays of dawn are breaking over the city. And, as could be said of any city in the world, this early morning light held the power to lighten the gloomiest heart and herald, with increasing hope, a brave new day... Not so for Harry, not this morning. 'Airport, quick as you can,' he barks at the driver as he enters a yellow cab, 'Big tip if you do...*toute suite.*'

The cab leaps away with squeal of tires before Harry has barely shut the door. The old city streets and magnificent buildings splay out and merge with Harry's reflection in the cab window, he looks at neither. Only when the airport looms does he pay any attention. 'Interstate, quick as you can,' he growls.

The cabbie takes a couple of turns, a couple of reckless passes and they arrive. Harry takes the pistol from his belt, removes the shells and, holding it by the barrel with the cuff of his coat, wipes it clean with his handkerchief. The cab stops, Harry pays the fare, tips handsomely then offers the cabby the pistol, handle first.

'By-the-by,' says he, nonchalantly, 'I found this on your back seat... can't just leave it there... God knows who'll get in next. Hand it in to the authorities, there's a sport.' He then walks away, leaving the bewildered cabby holding the enormous silver pistol. Harry deposits the cartridges into an adjacent trash-bin then continues on into the terminal to meet Rose at the check-in.

Rose is not happy. She is standing, one hand across her middle and the other in the small of her back in classic pregnant stance – a look of thunder on her face.

Before Harry can kiss her hello she turns on him coldly. 'You're still going on with it, aren't you?'

'Rose,' greets Harry, ignoring her bad mood, 'how are you, don't look so worried, it's going to be okay? By-the-by old girl, I've thought of a name, 'Barney' short for Barnaby, my uncle's name.'

'It's rubbish. And it still might be a girl.'

'No, it's definitely a boy... it runs in the family.' He attempts to kiss her again.

'Well, girls run in mine. Drop it Harry, for Christ sake drop it. Don't you realize, they'll follow you, they'll kill you, us, the baby.'

'I can't, Rose,' says Harry pulling back from his rejected kiss, 'I need it more than ever now... insurance, don't you see?'

It's obvious by the look on her face that Rose doesn't want to see. She turns and walks toward the check-in. Harry shrugs and loads their luggage onto a trolley, then follows after her.

On board Concorde, Harry offers to help Rose off with her jacket. She shrugs refusal and sits awkwardly in the window-seat, hands together across her stomach. Harry sits beside her and fiddles with her special, pregnancy seatbelt, attempting to fit it around her huge, swollen breasts. She gives him a filthy, 'are you having a good time' look. Harry eventually secures it then turns away.

The aircraft takes off. As the elegant craft skims effortlessly through the sound barrier Harry sits quietly working on his papers, trying his best to ignore Rose's brooding anger. Throughout the three hour flight there is little conversation, just a curt yes and no to Harry's inquiry as to food and drink, Rose, sometimes dozing, sometimes just gazing moodily out of the window, all the time looking angry and uncomfortably pregnant.

The order to 'secure seat-belts' flashes, prior to imminent landing. Harry tries again to cajole her.

'Everything okay, Rose... the baby?'

'What do you care?'

'Snap out of it Rose,' says Harry, rolling his eyes, 'You must see... I'd be forever looking over my shoulder. I know what that's like, I've led my whole life like that, waiting for the inevitable knock on the door. I don't want that for us. Those bastards owe me... owe us. They were going to kill me, I'm convinced of that... and that'll cost them. I'm on to something and–'

'Shut up. I don't want to hear. I've got other things on my mind.'

'You sure every thing's okay, Rose?'

Rose, with enormous effort forges a half-smile, 'Sorry Harry, I'll be okay. Just go back to your papers... you think more of them than you do of us.'

Harry shrugs and they fall to silence again.

CHAPTER FIFTEEN

Mercifully the aircraft lands. Harry and Rose make their way uneventfully through customs and on to arrivals. Harry leaves Rose and vaults the low barrier and runs up to Alfred, who is waiting to meet them. He grabs the little man up in a big hug.

'Alfie. God, am I glad to see you?'

'And me, you.'

'How have you been? Everything okay... you watered my mandrakes? All the telex and fax papers received and stacked?'

'Yes, yes an' yes... your garden is okay... What do you do for buttonholes over there?'

'Oh... plastic, it's only America, Alfie, mandrake doesn't bloom there like it does in England'

'Yes, well... nice bloody mess you're in, I see,' he eyes Rose's swollen stomach, 'Now put me down you silly sod, I'm with...' he indicates to a tall blowsy looking woman in her fifties across the big hall. She waves as Harry lets Alfred go.

'Very nice, Alfie, very nice indeed. Now, quickly before Rose comes out, have you done everything?'

'Yes.'

'Everything?'

'Yeees. Christ sake, when have I ever let you down?'

'Where's the zapper; have you got it with you?'

'Don't be bloody daft... you do know it's illegal in this country? It's in your bureau with the rest of your toys. By the way, we've had a message from an old friend.'

'The old man?' says Harry, grabbing him up again in his excitement,

'No, sorry Harry, the old man is still dead. It was–'

'God, I could do with my uncle now?' He looks to the heavens, 'Where the bloody hell are you?'

'It was–'

'Shush, later. Here comes Rose.'

As Rose makes her way over, Harry walks to meet her. 'Rose. I want you to meet Alfred and...' He raises a prompting eyebrow to Alfred.

'Ami,' says Alfred's lady-friend as she joins them, finishing the introduction, 'Hello Rose, I've heard a lot about you,' she gives Rose a curious once-over, 'Nice to meet you, I'm sure.'

'Right Alfie, we're off,' says Harry, eager to get away, 'My motor's kosher, I take it?'

'Of course, legally road-taxed and insured... for once.'

'And you've got yours here?'

'Of course... also taxed and insured, at my own expense.'

'And–'

'Will you bugger off... Christ, I'll see you in the morning.' Alfred turns to Rose, 'You realize what you've bitten off here, Luv?' Rose looks away, hands still clasped across her middle. Alfred looks worried. He turns back to Harry, 'She okay, Harry?'

'I'm in the doghouse, Alfie.'

'What's bleedin' new?'

Harry shrugs and they go their separate ways: Alfred and Ami – Harry and Rose.

The main room of Harry's apartment is stacked with

piles of exuded fax and telex sheets, plus various healthy looking pot-plants proffering little violet flowers. There is still every kind of computer and video equipment, ancient and modern, plus a bank of ten antiquated telex machines and a similar bank of fax machines, one of which has spilled paper out over the floor.

Harry and Rose enter. Rose flops onto the nearest sofa. Harry brings in the baggage, and then picks a fresh blossom from one of the flowerpots. 'God, that feels better,' say he as replaces the plastic prosthesis, 'I can face the world now.' He picks up the exuded paper and stacks it neatly on the mountainous pile of the same, then joins Rose on the sofa.

'It's the last batch,' sighs he with relief, 'I've got the bloody lot now. All the data from the first expedition, and all the data from your Mars shot, all in binary. Thank God no one thought to check the old-fashioned telex and fax lines – they're all so up themselves with the newfangled bloody e-mails and internet they hardly ever bother now.' Rose looks away, uninterested. 'Help me, Rose, I've got to scan every sheet. There must be two thousand, ten seconds a sheet. I've got two scanners, one hopper fed, the others hand fed… it'll take all night. But if you help…'

'If I must,' says Rose grudgingly conceding, 'but I'll need to rest first, unless you want me to start right now?'

'Sorry, Rose… No, you rest, I'll make us some dinner, then, when… if you feel okay we'll get started.'

Harry gets up and lifts Rose's legs up onto the sofa. Without a word she closes her eyes. Harry stands a while watching her beautiful face as she sleeps, then walks to the kitchen. After he makes dinner, he wakes Rose with a kiss.

As they eat, Harry describes what has to be done: 'What I have to do now, Rose, is scan everything then condense it and then put it onto Betamax... it'll take about twenty cassettes. Once I've got that I... I transfer it to Umatic, then I condense it again onto VHS, Then...'

'Then, what?'

'Then... Christ, I don't know, I fuff around all night... unless you do it, Rose.'

'Okay, you've almost got it – you're not quite as stupid as you look.'

Harry is relieved. Wanting to get started before Rose changes her mind, he leaves his meal and starts to feed in the sheets. The phone rings. Harry curses as he answers it, 'Damn. Alfie. What is it, what's up? ... Ami says what?'

'She says she's worried about Rose,' answers Alfred in ushered voice, 'she says her face don't look pregnant enough to match her belly... just has this feeling, silly cow. Anyway, I thought I'd share it with you.'

'Not now Alfie, Christ, not now. I'm bloody strapped for time, I'm desperate... in the morning, old luv, in the morning.' He puts the phone down and whispers to himself, 'Women, Christ all-bloody-mighty what are they like?' He considers for a moment, then goes to his bureau and fiddles, pockets something and walks over to Rose, smiling. 'Now this really has been in the family for years,' he twiddles a necklace in his fingers. 'Pearls and coral, from the South Seas... an engagement present... that's where I intend to take you for or honeymoon, the South Seas. I want us to start off properly.' He takes the beautiful necklace and offers it to her. He attempts to put it around her neck.

Rose turns away, 'Give it to me another time, Harry... 'I am not in the giving vein today'.'

Harry shrugs and puts the necklace back into his pocket. 'Ha, I like it... *Richard the Third* – Okay, 'another time.' —Let's get on.'

For the next six hours, they work together, loading papers into the scanners. Harry insists Rose take a rest. He leads her to the sofa. Once she is settled he returns to the machines and continues working, feverishly crashing buttons, cursing, and hurling spoilt efforts across the room. Rose, still half awake, lies watching.

At length, he turns to her, elated, with a final tape in his hand. 'Woo,hoo. Bloody Bingo.'

He fits the tape into the large video consul. Rose stands and walks to him and with one hand unbuttons her cardigan and lets it slip to the floor. Harry now squats in front of the console, fiddling with the controls. He runs the tape – just static appears. He adjusts it and runs it again. Cursing he adjusts and runs it again. This time, a picture starts to form.

'Got it,' he yells. 'First, the British 50's, Mandrake Expedition. Your US Junairo, Mars shot is still on the machine processing.'

Ghost-like images glide across the consul screen, computer constructed facsimiles of the real figures. They move around like animated airbrush drawings, suddenly deteriorating into shapeless confusion as holes appear in the data, then the computer-link reconstructs. The fuzzy-logic program assumes the data defect, fills in and automatically runs the tape back, then moves on. As other gaps appear, the computer program digests and becomes more acute. When the tape finishes, it automatically rewinds and starts to run again. On the screen the astronauts appear seated, then standing, now looking around in puzzlement, as if following the flight of a bird or insect, loose in the cabin.

WHAAAM! A hand suddenly crashes across Harry's face knocking him clear across the room. A table and chair disintegrate as he awkwardly glides through them. He immediately staggers up, bleeding from the nose and teeth. He looks back at Rose, standing calmly by the video. She takes out the tape with one hand – the other arm now hanging limply by her side and exposing the horrible scarring from her elbow to her wrist. Without need for words Harry backs up against the wall, his fist banging hopefully, instinctively, against the hard brick wall. He watches mesmerised as her good hand moves up to her neck and unbuttons her dress. It falls open to reveal her naked, bloated stomach. Harry's hand now feels for his holster.

'Looking for these, Harry?' It is Rosette that speaks. She picks up his holster from behind her chair and lets it drop.

'Rose, I–'

'I promise this won't take long.' She places her good hand onto her swollen stomach, pushes the fingers deep through the taut skin and rips her hand upwards, tearing a huge flap of flesh up across her body, up to the sternum, revealing a gorged blister of blood. The membrane ruptures and bursts open with a gurgling rush of matter, tentacles and steaming fluid. These tentacles whip at Harry's throat and body, grabbing and heaving him toward the glistening gash. Harry grabs at anything in reach for anchor. But he is slowly pulled nearer and nearer, until the top of his head is just inside the opening. With one hand he grabs at the slippery rim, trying to force his head out. His other hand is grabbing at something in his trouser pocket desperately trying to retrieve it. His head now totally engulfed as the drooling lips start to close.

His hand comes free from the pocket. He is holding a grey angular object in his fist. This he rams into the mess of blood-slime, and with one last almighty effort, forces his head free. A flash. Followed by a blood-curdling shriek, then a crash as Rosette is flung across the room, the slimy tentacles flaying out a fountain of bloody fluid with the force. Harry is winded and stunned from the partially conducted electric shock delivered from his 'zapper' – the police-issue stun gun Alfred had acquired.

Shaking furiously Harry manages to stagger to his feet. Hardly able to stand, he grabs hold of the video console for a few seconds to steady himself. He manages to gather his wits. Still dripping glutinous digestive slime he grabs up Rosette's discarded cardigan and wipes off the toxic goo from his face, neck and hands. He takes a few steps towards the convulsing heap, the tentacles flaying aimlessly in all directions. Bending down shakily, and making sure his feet are on dry carpet, he delivers another bolt. A harrowing cry as Rosette's stunned body leaps again. Harry stands and staggers to his holster where she had dropped it. He injects a full syringe into her heart, then sits back on his heels and watches the galloping pulse gradually die to a flicker. She gives a last erratic leap then is still. After a moment he injects the other syringe, refills and injects again. Satisfied that Rosette is dead, he covers the ugly mess of her lower body and stands back and looks down at her exposed face. 'What the hell have you done with Rose, you bitch?' he hisses into her staring sightless eyes.

After covering the rest of her he picks up the telephone. It rings for some minutes before Rose answers. 'Harry, is that you? Do you know what time it is? Hey! Who is this? Answer damn you.'

Harry is so relieved he can hardly speak, 'Rose... Thank God... Rosey... thank – '

'What's happened,' yells Rose, 'where are you? Pull yourself together, take deep breaths. Now, again Harry... Damn you, what's happened?'

'It's over, Rose, she's dead. Rosette is dead. Can't say too much, your phone is bugged... be careful... don't tell anybody. I'm coming back for you... be ready... I love you... your phone is bugged.' He finishes his rambling and puts the phone down without noticing the curtain blowing in from the open window and the shadow that momentarily flits across the far wall.

He makes another call, to Alfred. He waits a moment and the line connects. 'Alfie, tell Ami she's a bloody genius. Now, I need you to book a flight back to the States, New York, first Concorde available, just for one. Let me know when you've done it. No no no. Do not come here. I'll give you instructions when you call back. As quick as you can, old mate, I'm bloody desperate.'

Harry puts the phone down and goes into the bathroom, picks up a remote control and turns on the shower, strips and showers. No gargled Jerusalem, just a silent prayer. He dries himself and puts on his dressing gown.

A noise.

Harry is frozen to the spot with fear... only his eyes are mobile. Something in the next room is moving. He slowly picks up the remote and flicks off all the lights. – Silence. He waits motionless. A creek of floorboard. In the gloom he fumbles in his pocket for his stun gun, his hand comes out empty. He creeps slowly through the apartment making for the main door, the noise following close behind.

A horrible, coarse, clammy, hideous inhuman hand

slithers along the wall behind him. Harry is near the door and about to flee. The monstrous hand catches him around the throat mid-step. Harry leaps about a foot in the air and out of the hand's grasp, only to be grabbed two-handed on his descent.

'Got ye, yoo bustard! A whole fockun' weekend I spent in that poxy lift, sitting in my own shite.'

'YOU.' screams Harry.

'Yes, meee! Yoo should'ni leave your windies open, Laddie, not with scum like me around. Now, you poncey sassenach poofter, I want my money, plus interest. But first I'm going to bust this wee bottle inti your pretty face.'

Harry, surprised, relieved, and now angered, smashes a knee into Radcliff's undefended groin, then delivers a masterly head-butt to the descending nose followed by two inspired uppercuts, left and then right, from somewhere just below his waist, punching clear through towards the ceiling, finishing with a beautifully aimed drop-kick to the throat of the now doubled-up Scotsman, sending him sprawling head over heels into a face-down backside-up heap.

After a moment of reflection, Harry finds his syringe and delivers a short, sharp squirt into the offered rump. He then grabs his would-be assailant, shoves a more than ample bundle of bank notes into his pocket and then hurls him roughly out of the door, down the stone steps and out into the cobbled street, adding a parting comment, 'Sorry, old luv, but you picked an inopportune moment. Now, you've got your money so I'm sure we won't need to discuss this matter further.'

After a few moments, Radcliff staggers up, looks back at Harry in wild disbelief, and then hobbles off, one leg immobilized and continually floundering under him. The door slams.

CHAPTER SIXTEEN

Having sat brooding the four hours it took Harry to drive from Kenney Airport, Rose now sits expectantly on the stoop of her mother's house. When his car finally pulls into the drive she runs to meet him. As he bundles out of the car she grabs him up in her arms, almost knocking the video recorder and case out of his hands. She holds him to her until Mabel comes running out.

With four arms tightly around him, he is bustled into the house. Once inside he is sat down and interrogated in tandem, one from the other. Mabel finally relents and, after making Harry a meal, retires to bed leaving them alone. The minute she's gone, Rose starts on him again.

'My God, Harry, you went all the way to Edinburgh with that–'

Harry holds up his hand. 'First things first, Rose. I've got a present for you. I hope you'll accept it this time.' He holds out the necklace he'd offered to Rosette.

'This time? You don't mean to tell me you offered it to that freak?'

'What? – No no no, of course not. It's an engagement present. I want to do things properly.'

'You did, didn't you? How the hell could you have mistaken her for me?'

Harry looks away, sheepishly 'I knew it was her, I was just–'

'Liar,' she says it as she offers her neck. Harry fixes the clasp.'

'I, I knew all along. I was just playing–'

Her eyes again say 'liar'… then she smiles forgivingly and touches the necklace, 'How long has this been in the family?' says she, mocking, 'It's beautiful.'

'It's tear-drop pearls and coral… I had it made to match your eyes.'

She punches his arm. 'That's crying over you, Creep. My god, me crying over a man… what a switch.'

'I'm not just a man, Rose.'

'No. Not 'just a man'. So what now, do I take this as a proposal Harry?'

He hesitates, she puts her clenched fist under his chin in a mock threat.

'I love you Rose, but if I marry I inherit a fortune. You may not like me when I've got a fortune.'

'I'll force myself, we all have to make sacrifices.' She pushes her fist further into his face, forcing him off his chair and onto his knees.

'Having money may change my whole character.'

'What? You haven't got any 'character.' On your knees, creep, your last chance.' She falls on him, kissing him as they tumble.

Harry pulls away. 'The baby, Rose, mind the baby.' He picks her up in his arms and carries her awkwardly to the divan. Rose giggles at his feeble effort as he all but collapses onto it. They sit quietly for a few moments.

'You should know,' says Harry, breaking the silence. 'I'm going to write a thesis. Now it's over I want to get it all out of my head and onto paper, then I can leave it. Will you help?'

'I'd rather not. It still makes my teeth shiver. I don't even want to think about it.'

'I think we owe it to my uncle, Rose,' says Harry,

trying another tack. 'I'm writing this account and having it bound with his. I'm calling it The Mandrake Syndrome – It's okay, I'm not going to publish it, it's for the archives.'

'Okay,' says Rose, grudgingly, 'Then it's done with, yeah?'

'Yes yes, definitely: I want you to tell me, first hand, your feelings when you first saw your double.'

'You really want to do this? Can't you just read the transcript?'

'You are bloody joking? I'm not going back to get that. You think I'm crazy?'

'Okay, calm down Harry. I have a copy, here.'

Harry sits up and diligently takes out a notebook and pen from his case. 'Anyway, I don't want the transcript version, I want your version.'

Rose studies him with contempt. 'Surely you're not going to do this now?'

Harry ignores. 'What was your first reaction, Rose, your first thought? That's what I want... that's not in the transcript... when you looked into Cowen's coffer and saw you were looking at yourself.'

'Oh, God, I don't know,' says Rose, reluctantly, 'what would you think? – Hey, wait a minute...' she stops and thinks a moment. 'What do you mean, 'when I looked into Cowen's coffer?' I was still in my coffer, she looked down on me... she was the first thing I saw when I opened my eyes – the goddam freak.'

'The tapes, Rose,' says Harry, now with hint of authority in his voice, 'Rosette destroyed them, but my tape machine automatically made a duplicate copy. I've seen it, but I don't understand it.'

'Seen what, for Christ's sake? Seen what?'

'I couldn't quite make out what happened. When the

thing entered the Junairo, just after you rounded the planet, it seems the men are fighting something... lots of motion. Then it's unclear... lots of thrashing about as if something is loose in the cabin. Then... well, I couldn't keep it... we lose it. The next part is of Junairo closing on EarthlabOne, just before they make audio contact. You appear to be the only standing figure.'

'God sake, stop.' says Rose, now on the verge of tears, 'I don't want to hear any more.'

'You have to, Rose, you have to. You were just standing looking down into Cowen's coffer. When the figure emerges from it, it is Rosette in Cowen's suit... you can see the name clearly. Now here's the part I don't understand, you don't seem shocked. You just walk over to the lab section and pick up the lance. Rosette stands there, almost casual.'

'Stop it, Hal. I can't believe it. My God, she must have had some kind of control over my mind.' She stops and puts her hand protectively on her stomach.

Harry gives a despondent sigh. 'It's true, Rose, it's there on the tape.'

'I won't believe it until I've seen it. Run it Harry, run the damned thing.'

Harry picks up the remote control and walks to the video player and slips the tape in.

'What else did that freak make me do?' She screams from across the room. 'No, don't run it. I don't want to see it.' She runs to him and holds him. Harry, still holding the remote control, presses the play button in the direction of the video recorder. 'Hal, for God sake let's leave it, it's going to drive me crazy.'

'It's okay, Rose, it's over. I told you, it's all over–'

She kisses him, muffling his words with her lips. He is

lost for a moment in the fragrance of her hair and the warmth of her body. The whole dreadful mess seems to ebb away under her flood of honey sweetness.

The video constructs a picture. They both turn and watch: A ghost-like Rosette starts to climb out of the coffer. Rose is just standing, watching and waiting. Once out of the coffer, Rosette nods and Rose picks up the lance. After a few moments, Rose cries out from the video. 'Oh, my, God. I must be dreaming. Can you see this, Major?'

Rose turns away from the screen. 'For God's sake, leave it, Harry... it won't make any difference. Whatever happened out there, I still love you.'

'No!' he yells, pushing her away from him, 'Get away from me.'

Rose is shocked by his outburst. He holds her at arms length then pulls her towards him as if in a last embrace. With all his strength hurls her across the room. She crashes to the floor. In an instant he has his syringe in his hand, the other hand still holding the remote. He starts across the room toward her. The door opens just in front of him and Mabel enters and stands between them. 'What in heck's going on?'

Harry brushes her aside and lunges toward Rose again. Rose is just about up. Mabel recoils and leaps at Harry's back. A long serpentine tongue whips out around his neck. The vice-like grip needs no explanation. Harry lets loose a reaction right-hook that disappears over his left shoulder, the syringe held tightly in his fist. The razor sharp needlepoint finds the white of the old lady's eye – novocaine voids directly into her brain. Her grip slackens and she slithers to the floor in convulsion, spewing out blooded mucus. Harry immediately grabs his other syringe from inside his jacket.

Rose now stands facing him. They stare across the room at each other, Rose in typical heavily pregnant stance, one hand on her stomach the other hand in the small of her back.

'You won't use that Harry, you'll harm the baby.'

'You mean there is a baby? I've seen your babies.'

'There is a baby, Hal.'

'I don't need to use this, Rose. I told you, it's over. There was always something nagging at me... Why was all the flesh from Cowen and Fitzgerald used up?'

'Does it matter?'

'My uncle's calculations were right. He said it uses one body as material and just over half another body as, shall we say, fuel. That's why we always found a remaining torso. Two bodies make one, with half a body left over, but three bodies make two, with nothing left over... elementary physics really. I checked with the old man's calculations, pound-for-pound, what was made, what was lost. You couldn't have made Rose and one of the men, there wasn't enough material. But there was enough for two of you – just enough.'

Rose takes a step towards him, 'It won't make any difference Hal, I love you and you love me, Rose. So I'm half Martian,' she smiles sardonically, 'Don't tell me you're *speciphobic*, Harry? New word, you heard it first here.'

'Yes, that is funny.'

'See, gods can be funny, Hal.'

'I'll laugh later, when I've got less time?'

'I really do love you, Harry.'

'You were the smartest on that ship, Rose, and when it came to a fight for intellectual superiority there was no contest, you won hands down. You worked it out; you needed a scapegoat. The whole bloody thing was a sham right from the start, even Rosette's rogue cell disorder, all to throw them off the scent. And the only danger was myself. You knew about me, and my uncle's records.'

'Yes.'

'Why didn't you just kill me, you had enough chances?'

'I had to run with it. Someone else would have taken your place... someone competent.'

'Oh, thanks very much.'

'Anyway, Harry, I fell in love with you. I'm still human... just a little different that's all. It really is your baby, yours and Rose's... mine. You've seen the ultra-scan... you've even named him... Barney. It will still work, Hal, you'll see. You'll have knowledge.'

'Knowledge?'

'Yes, knowledge.'

'Will I be like Kit Marlowe's *Faustus*, selling my soul?'

Rose ignores. 'This isn't our first time, Hal, we've been here before, thousands of years ago, before your recorded history. When we first came you were Neanderthal, when we left you were Cro-Magnon. That was our legacy – we are in your legends. '

'Is that what you are, Rose, a god?'

'If you like. I'll make you a god too, Harry, a Titan.

'And which one are you?'

'Oh, believe me, Harry, you'll know me when you see me.'

'The baby, is it one of your kind, will he be a god?' Harry raises the syringe ready. Rose backs away.

'I told you, Hal. We don't breed, we're parasitic. You

know what we do. But the child will be human.'

'I don't believe you.'

'Yes, you do. I wouldn't lie about that, I love you. It'll be in the child's interest. You're a flawed race... you'll eventually destroy the Earth out of greed and selfishness... it's inevitable, even you must see that. It's the only place, Harry. There's nowhere else in the Galaxy. The only life is here, and we've as much right to it as you, more, we don't covet, we don't despoil, we don't destroy. We're more deserving, more human, than you are.'

'What about Mabel? You don't seem much upset about your mother. You're not human. Who did you kill for her donor?'

'I killed no one. Cameron sacrificed himself, it's what he wanted, for...'

'For... what? For, what?'

'For the Greater Good, Harry. The salvation of our species, ours and yours.'

'What utter, utter rubbish.'

'Well... if that's your final word?'

'Yes.'

'You'll never make it – you do know that? I promise it won't take long. I really do love you, Harry, and I really was jealous of Kate. And I promise I'll look after our baby.'

Harry takes a step backwards. Rose rolls her eyes, her irises disappearing under her lids. Her mouth sags open and starts to drool, wider and wider. Her mouth is now a gaping gash, awash with thickening, digestive slime. From this gurgling mess, the tongues leap: three, six, and now a dozen. Her head tips back grotesquely and more blood-coloured tongues spill out, thrashing, whipping. The Medusa stands trembling like a great sea anemone, weaving its snake hair, as if through a rushing current.

Harry's right hand drops to his side in shock. The heavy syringe slips from his slackened grip to the floor. The Rose-beast takes a lumbering step toward him. Harry lifts his left hand to shield his face, the remote-control-unit still clasped in his sweating, white-knuckled fist. Nearer it lurches, its head tipped back, just neck and tentacles surrounded, in comic horror, by the beautiful pearl and coral necklace. Closer and closer comes the gaping mouth and necklace – closer and closer come the cluster of pearl and coral-coloured pellets – the same pellets as the ones Harry had fixed to the lift device.

He pushes the button on the remote. A blinding flash accompanied by an ear-splitting crack, like the noise of water spilling into boiling fat as the ring of thirty pellets detonate, hurling the Medusa head, tentacles, and a huge gout of blooded strands, to every corner of the room. Harry simultaneously falling to the floor and shielding himself as best he can. After a few moments, he staggers up and edges along the wall and out into the hall. He wipes away the blood and tissue from his face and hands and stands with his back to the wall, not daring to look back. A long silence, then he reacts to a faint rustling from within the room. His eyes widen with fear. A gurgling of fluids being displaced, then a pitiful cry, no more than a whimper. A look of horror now occupies Harry's face as he enters the room.

What is left of Rose is piled in the middle of a widening pool of blood, her legs buckled under her as she had fallen. Something is moving inside her sodden dress. Harry picks up the syringe and puts it between his teeth. He bends down and slowly thrusts both hands into the grisly pile of steaming offal.

Bewildered he lifts a blood-soaked baby, umbilical still attached, from the gory mess —a perfect boy-child, wailing for its demon mother. Harry blinks unbelievingly at his son.

> *And all should cry, Beware. Beware.*
> *His flashing eyes, his floating hair.*
> *Weave a circle round him thrice,*
> *And close your eyes in holy dread,*
> *For he on honeydew hath fed,*
> *And drunk the milk of paradise.* Coleridge

END BLOC ONE

BLOC TWO – SILVERCORD

Author's note:

I feel obliged to offer this short testimony – not wishing to teach any astral-projecting grandmother to suck eggs – concerning the phenomena *The Silver Cord* and *The Golden Bowl*.

The Silver Cord, as we all know, is the metaphysical ligature; a term referring to the connection between the physical body and the astral body. The Golden Bowl, this being the Astral Body or the Higher Self. During astral projection, the mind leaving the flesh-and-blood body, one can see, as some projectors claim, a silver cord linking the astral-form to the physical body. This cord mainly appears to the layman projector as security that they will not become lost. However, even experienced projectors find it useful, claiming it as the preferred way to return to the physical body.

The said phenomena are derived from the Bible: Ecclesiastes 12. 6-7: *"Remember him, before the silver cord is severed, or the golden bowl is broken; before the pitcher is shattered at the spring, or the wheel broken at the well, and the dust returns to the ground it came from, and the spirit returns to God who gave it."* Occultists and mystics, especially in contexts with Bi-location, Remote Viewing, and Near-death Experience, mention the Silver Cord. —Please read on…

1983: Two years on.

CHAPTER ONE

A swirling mass of gas and matter, a chaos of blood and tissue mixed with sound of men screaming, rising to crescendo. Then, like a hammer blow, silence.

The blood-mist begins to clear and through the dispersing vapour a spacecraft cabin appears. At its centre is a wheel-like structure bearing four life-support coffers. Encased inside one transparent dome is the inert body of a man. Through the condensing mist, the features of Harry Mandrake gradually sharpen into focus. This coffer is linked, toe-to-toe, with a partner coffer on a centrifugal turntable. This second coffer is breached and filled with vapour.

The mist clears a little more. Now visible below the first pair of coffers are two identical coffers, forming a second layer. They are held apart in opposed juxtaposition to the top pair by a huge spindle. One of these lower coffers is also smashed, the other is covered with blood and crystal fragments.

The mist is now gone and the blood-spattered cabin of the United States starship *Junairo* comes clearly into view, but with no living eye to observe it. The plain, once orderly cabin is strewn with the remnants of imploded bodies, unrecognisable in number, strands of bloody tissue sucked into every corner of the cabin by the ravenous vacuum. Silent flashes of an arc-welder accompany the shower of molten-metal droplets that dance lazily in the manufactured quarter Earth's gravity. The ship's SAMS, *Self Activating Maintenance Systems*, scurry to repair the breach in the triple skinned shell of the bulkhead. These are the closest thing to life on this otherwise deserted spacecraft. Ungoverned by human hand the small spider-like drones crawl over the craft calculating with their artificial logic, and repairing the damaged hull.

One of the SAMS crawls through the jagged hole and out into space, the same gruesome route as the pulped crew. Here it will repair and close off from the outside, thus condemning itself to an eternity lost in space. It finishes its noble task and drifts away from the bulkhead of USS Junairo2, a sleek pencil-thin compilation of metal cylinders. To the stern of the craft is a cluster of iron-fusion engines massed together in a cluster, making the whole look like a gigantic silver sledgehammer. Emitting from this massive power stack is a vaporous veil slipstream a hundred times the size of the ship, forced out almost to right angle. From its centre is a white-hot nucleus of ionised coherent material throwing out to infinity.

A snow-covered forest – ALASKA, 1983: A distant figure of a man crashes through the sparse undergrowth. He is dressed in a light reflective suit and what appears to be a grey service balaclava with huge, head hugging snow-goggles. He is pursued by a pack of hungry wolves.

The pack led by a huge grizzled male now encircled the man, who drops to his knees awaiting the attack. The pack leader approaches with bared teeth and hurls itself at the man. In a savage flurry of tooth and claw, it is the wolf that is ripped to shreds. Now the rest of the pack descends. The man tips back his head and howls up at the angry sky. He is blooded around his head, and around the snow-goggles his service balaclava is all but ripped away. The wolf pack checks its advance, and then in unison join the victory-cry in recognition and homage to their new leader.

The man grabs up the butchered wolf and rips into it with his teeth as he runs, feeding on his feet. The subordinate pack follow, taking his lead into the snowy wilderness.

CHAPTER TWO

Harry Mandrake—forlorn, desolate, the past two long years incarcerated in the Carnegie Space Agency hospital as high-security inmate—has a visitor. It is his one-time friend, Hamish, now slightly running to fat that greets him as if still old buddies.

'Harry... Hal. Good to see you. It's been a long time. You're looking older. Here... they let me bring your lunch.' Hamish, carrying a dining tray full of food, is sporting an insincere, manufactured smile. Harry snarls and takes the tray. Hamish studies the weary face before him. Worry lines and a touch of greying hair to the temples now contend with Harry's boyish looks.

'You want something, Fatman?' says Harry, totally unwarmed by Hamish's greeting.

'Hey. Can't I come to see my old friend without my wanting something?'

Harry ignores the rhetorical question. 'What ever it is, I first want to know what's happened to my son... to Barney? Whatever you want from me, my price-tag includes Barney.'

'Price,' says Hamish, mimicking hurt feelings, 'You still on the make, Harry? Nothing never changes with you, does it?'

'You bastard, I haven't seen my son for two whole years.'

'Hal, believe me, he's okay. He's passed every test they can think of, and a few more.'

'You can't do this. I must–'

'Must?—What are we going to do about you, Harry?'

'Go to hell.'

Hamish smiles and moves closer. He hisses his words with cold, calculated hatred into Harry's ear. 'You killed three of our people, you gutless fucking wonder. If they'd let me have my way I'd have wasted you back then, in New Orleans.'

Harry pulls back a few feet and with all his might head-butts Hamish, then hurls the tray and follows after the descending man grabbing at his throat, yelling, 'You bastard, I'll kill you... I'll kill you.'

In a fraction of a second Hamish is up on his feet and holding Harry, still screaming abuse and clawing at the air, at arms length.

'Tch, tch, Mr. Mandrake,' says Hamish, ignoring the little trickle of blood from his forehead, 'such language. What would little Barney say?'

'Go to hell.'

'I'm sure he wouldn't say that.' He pulls Harry nearer. 'Listen you pumped-up Limey faggot. I never liked you. You murdered Rex. You suckered him with friendship... that won't work with me. He was worth ten of you. It's not over, creep. *The race is not over, it goes on, Judas.*'

'If you are going to use a quotation, you ignorant moron,' says Harry, speaking with difficulty, 'Get the damn thing right, 'It goes on... Judah.' – *Ben Hur.*'

Hamish smiles, 'Well hush my mouth. Now, old man, are you going to behave yourself?' He lets Harry go.

'Okay,' gasps Harry, rubbing his throat, 'you want something or you wouldn't be here. Well, I want something too. I knew you'd be back. It's still bloodywell out there, isn't it... And you're here to deal?'

'Maybe.'

'Bull,shit. I'm prepared, see? I've been prepared the whole time I've been in this filthy, horrible place. I want money, lots of it, and I want my son, and I want out of this God-forsaken country. Nothing less.'

'To deal?' says Hamish, his mood now changed to one of indifference. 'No, Hal, not to deal, to beg. Major sent me to beg if necessary, 'offer the Limey fuck anything.'... his words.'

'Well, you tell Major he can go stuff himself.'

'*Stuff himself?* – I'll put that to him. Now, I'm laying everything on the table, Hal, because...'

'Lay what you like, where you like. Do I look like I give a damn?'

'... because I want you suspicious. I want you scared fucking shitless. You won't have any trouble with that.'

'And if I refuse?'

'Oh, I want you to refuse. I want you banged up here till you rot, you gutless fucking coward. I want – Aghhhhhh.'

Hamish recoils backwards, lifting his hands to the stabbing pain in his shin, of which Harry has just viciously kicked.

'You bast–' his expletive is cut short by an equally vicious kick to the other shin, 'Aghhhhhhh. Jezuuus!'

Unable to resist the instinct to lift his other leg, Hamish crashes to the floor into the scattered debris of the dining tray.

Harry leers over him, 'I may be in a nursing home, old sport, but that doesn't mean I'm bloody senile, nor that I'm wearing slippers.'

Hamish is immediately on his feet and hurling a punch. Which is interrupted by a crunching SMACK! The male nurse who, having just flung open the heavy oak door in response to the commotion, stands agape as the edge of the said door absorbs the said punch destined for Harry's nose. The oak door resists, undamaged.

In his office, Major sits calmly at his desk reading notes. Hamish stands to his right as sentinel. The big man's hand is taped to a multi-fingered splint and resting uselessly in a sling. Standing the other side of Major, in the place Rex used to occupy, is a huge six-foot plus, beautifully muscular young woman. Major looks up from his papers as Harry enters escorted by two aids.

'Mandyke... Come in. Sit down. You know Hamish.' Hamish stares icily. 'And this is Rensa Lansavich.'

'She's Russian, Harry,' adds Hamish. 'She's here on behalf of her country to observe. No more secrets. Try not to kill her, old sport.'

Harry gives Hamish an equally icy stare.

Major gives an indulgent nod to Hamish and continues. 'The deal is the same as before, Harry. If you're in, you're in all the way, no turning back. Now I'll–'

'Cut the crap,' spits Harry, 'You want something, I want something. Let's trade.'

Major shrugs, 'Okay. What do you want from us?'

'I already told this,' he indicates to Hamish, 'gormless pillock. I want money... lots of it, and I want my son, and I want out of this damn God-awful, fascist bloody country... both of us... nothing less. Now, what do *you* want?'

'Very eloquently put. Yes, we want something from you. Trust me, Harry, America and the whole of mankind want something from you–'

'Trust America,' gasps Harry, 'Not bloody likely. As

my illustrious uncle used to say, 'never trust a country that doesn't play cricket.' And as I say, 'cut the crap.' What do you want?'

Major shrugs again. He looks at Hamish, then back to Harry, 'Information, just information. Some of your uncle's papers don't quite make sense. We need you to clarify.'

'Is that all?'

'Hopefully, yes. Are you familiar with the SilverCord project?'

Harry's jaw drops in shock at this revelation. He recovers and hurls his words back at Major. 'How the hell did you get your hands on that? That's still under embargo... even I can't get access to that.'

'Calm down, Harry,' says Major with an ironic smile. 'You're forgetting you are a convicted felon. You don't have access to the air you breathe unless I say so. Since your murderous little episode your pink-pussy government has been obliged to lift that embargo; in the circumstances, it was the least they could do. Now we have everything... that's the real everything, not the 'everything' that you promised us in the beginning of our, shall we say, coalition.'

Harry calms and sits defeated, putting his head into his hands. 'If you've got everything, you don't need me. And for the record, I've been convicted of nothing.'

Hamish leans towards Harry, spits his words with relish, 'You were found unfit to plead. In your absence, you were adjudged guilty but with diminished responsibility... of murder and various other crimes. Subsequently, you have been committed to this sanatorium on, as you Limeys say, 'her Majesty's pleasure', vis-à-vis, Major's pleasure.'

Major holds his hand up to Hamish, as to say 'enough.' He turns back to Harry. 'And yes, Harry, we do need you… desperately. You are the only one who can help us. The shield failed and we lost an entire crew.' He studies Harry's face as he continues. 'We jettisoned what was left of them in deep space.' I take it you've heard?'

'No. I've heard nothing for two years. Goddam it, I told you it wouldn't work, didn't I? Well, <u>didn't</u> I? —My God, talk about 'once more unto the breach.' Jesus Christ. A whole crew?'

'Yes, a whole crew – Okay okay, you told us.'

'But you went ahead anyway. And you call me murderer.'

'Okay. We don't want… we can't risk anybody else. But we are contractually committed to sending a manned shot to Mars or we're dead… the Agency, that is.'

'But you can't. You can't.'

'Yes, we can. We have to. Look, there's a rider on the US Government deal, a factor figure, our profits are determined by it, and we have a shortfall. We need ass-on-planet in the next six months or the Agency is finished.'

'So what? What do I care for your dammed pox-ridden Agency?' Harry looks into Major's face with cold accusing hatred. 'Rose cared for the Agency, and–'

'And… you fucking killed her… twice. You 'put out the light, and then put out the fucking light.'… I know Shakespeare too, Hal.'

Harry leaps from his chair, his hands aimed at Major's throat. Rensa, who till now has stood passively listening vaults the big desk and has Harry in a full nelson before he's taken a couple of steps.

As she holds him she whispers softly into his ear, in perfect American accent, 'Hold on Mr. Mandrake, don't let yourself be goaded, that's what they want... please,' she holds him until he calms, 'Please, Henry.'

Harry calms and she releases him. He looks up at her. 'Don't call me, Henry... Harry, if you must.'

She smiles then takes her place at Major's side. Harry takes his seat again.

Major continues. 'Right, now you know not to mess with me.'

'Yes, or you'll set your big sister onto me.'

Major ignores. 'In, or out? If you don't help you won't see your son until you're an old man, and he'll never know about you. He'll be adopted and I doubt if you'll ever find him. And that's legal.'

'You can't–'

'Yes, I can. It's the law in this 'God-awful country,' old man. Just remember, you're criminally insane until I say different. Harry sits again, bravado gone, and puts his head back in his hands.

'He's all I've got left of Rose... don't you see? – Okay... what choice have I got?'

'Not a choice in hell. Now...' Major takes out a battered pink file and lays it on the desk. Harry looks at it, scarcely believing his eyes.

'Dear God in heaven, I haven't seen that since I was sixteen.'

'The SilverCord Effect,' continues Major, 'It doesn't work.'

'It bloodywell does, I've done it.'

'Get your coat on, Mandyke, we're going for a little ride. I want to show you something. Be prepared for a shock.'

'And try to keep a tight ass,' says Hamish, adding a contemptuous grin.

CHAPTER THREE

A glorious sunrise-halo hung over the cluster of dismal hangers and the control-buildings of *Camp Hero*, the Carnegie Agency airfield installation, Long Island. The morning's crimson blaze illuminating the buildings and the speeding vehicle hurling down the disused runway with a sanguine hue of blood: An eerie archaic predict of that which is about to come?

The group of mesmerised spectators – two guards, Major and Hamish, and Harry, dwarfed by the huge but perfectly formed Rensa – stand studying the speeding vehicle as it zigzags through a gauntlet of bollards. A squeal of breaks, a cloud of burnt rubber, and a momentary loss of traction, has the vehicle completely about-faced, set to continue back down the strip without loss of engine revs. Again the vehicle negotiates the difficult course, again at high speed. This time purposely driven into every bollard, splaying them out like bowling-pins and hurling them tumbling into the air.

The vehicle now levels with the group. The driver wittingly spins the steering wheel and the car veers sideways. It somersaults onto the roof, its side and then back onto its roof and continues on, sending out a fountain of sparks, glass and metal debris as it skids and pirouettes to destruction. Finally the car lands in a heap, right-side up in a cloud of steam and high-octane vapour.

'Christ,' yells Harry, 'we've got to get him out.' No one responds. He yells again, 'Move. What's the matter with you all?'

Major grabs his arm, 'Cool it, Hal… just watch.'

A few agonising moments pass, then the door of the flaming wreckage kicks off its hinges with tremendous force, tearing off half the battered front wing with it. Harry looks on in disbelief as a huge, grey-faced figure emerges from the now-burning vehicle. He is dressed in a shiny foil suit, a service balaclava and tight fitting goggles. Gracefully, in spite of its great size, the figure starts to run towards them. It takes just a half-dozen strides before the car explodes. The plume of fire throws upwards and outwards completely engulfing the grey figure, and creating a huge shockwave.

Harry holds his nose and blows out his eardrums. He turns on the group with disgust. 'You callous bastards. We could have got him out.'

Major grabs Harry's arm again, 'Watch, Harry. Just watch.'

From the ball of flame the figure emerges. It runs from the inferno hitting at its grey, naked flesh, the foil suit all but burned away. Once clear of the heat-pall the grey man falls to his knees.

Harry pulls away from Major's grip, leaps the barrier and runs to the kneeling man. He stops short. 'Dear God.' he gasps in shock. He is looking through the remnant of the service balaclava and into the face of a manufactured creature, its grey plastic skin blistered and melting, hanging off his huge exposed metal cranium. In place of eyes are what appear to be snow-goggles, set directly into the grey, gargoyle head.

The grey man looks up at Harry and extends its smouldering hand in greeting, and through blistering lips smiles, 'Mandrake, I presume? Alan Redman, at your service.'

As if the strings of life were severed in a single

instance, the grey man keels over, lifeless.

Harry bends down and lifts a huge grey hand, tries for a pulse. He turns, and calls back to the group, 'Quick. There's a slight pulse. – My God, there are two pulses.'

Major joins Harry, kneeling by the body. Harry looks up at Major with a mixture of contempt and puzzlement in his eyes, 'They've stopped. What the hell was he?'

Major smiles, 'Carbonised mesh fused with animal tissue, lab propagated. It is an amalgam of muscle grown directly onto the titanium chassis/ skeleton. We use the muscle, digestive system and nervous system from various canine donors, mainly wolf – the brain. The rest is mostly the Inuit husky breed, for obvious reasons; Mars is, like the song says, cold as ice.'

'Grown?' says Harry, amazed,

'Yes, it's a whole lot of medical technology come together. They've been growing skin on mesh since World War Two. Skin grafting, and organ transplanting, and microsurgery is, today, all but redundant. It is now replaced with chimeric-neurosynthesis, chromomorphosis, stem-cell HLA antigens, enzyme manipulation, RNA and DNA hybri-morphosis – Great words, yeah? I know it's difficult to understand because I don't understand it myself. But don't worry, I know a man that does… you met him earlier.' He smiles at Harry's bewilderment, 'Don't worry, Mandyke, all will be revealed.'

Harry, still holding the grey hand, shakes his head to make sure he's not dreaming, 'Why two pulses, Major?'

'Simple. Two hearts are more efficient than one… back up – twin carbs. And anyway, the blood and lymph system of the husky and the wolf are too small. There are lots of duplicated functions like this.'

'But it was human, you bastard. It was alive. He had a

name, Redman... God's sake, it spoke to me.'

'Yes, Redman in proxy. He was alive, and still is... whether he's human or not I've had my doubts for some time. But judge for yourself, you'll be meeting him soon. If we can't go to the stars in body, we'll go in spirit. Get a good night's sleep, we start early here.'

Harry lets the great grey hand flop to the tarmac.

The Carnegie Space Agency laboratory building, Camp Hero, is sectioned off into small atmospherically sealed glass cubicles. Inside one, Harry now holds a duplicate grey hand. He, Major and Rensa, all dressed in transparent overalls with dome breathing-filters, are being shown around the complex. In each cubical, a grey synthetic cadaver lays inert, some of which are partly obscured by linen cages covering arm, leg, and torso, some naked. Teams of similar-suited medics busy monitoring the peripheral, attached to each synthetic by pipes tubes and wires.

A man enters the laboratory and joins the group. He is small but hypo-energetic with the balding remnants of ginger hair and a close-cropped copper-collared beard. He is wearing a bomber style tunic in matching colour, and has a sardonic, impudent look on his face. As he approaches the group, Major introduces him.

'Hal, this is Alan Redman. You've already met.'

Redman winks at Harry, 'Mandrake, I presume.'

Harry recognises the Scottish-American voice as that of the deceased grey man. 'How'd you do?' says Harry, suspiciously.

Redman offers his hand as before. 'So, so... I do okay I guess, but to the point, how do *you* do?'

Harry, perplexed but still insolent, cautiously shakes

the hand. 'I'm of the opinion that I might be on the verge of feeling just an itsey-witsey bit better... maybe. But don't quote me, old man.'

'I wouldn't dream of quoting you, old man,' says Redman, equally insolent.

There is a noticeable air of hostility between the two, so much so that Major feels compelled to step between them.

'Okay okay. Now, Harry, you've got half an hour to ask these guys questions. Ask anything you want, but,' he points to the medics, 'be on your guard, some of these are G-men, government people. For god's sake don't let slip we have a problem.'

Redman pokes his impudent face close into Harry's space. 'Don't you worry your sweet sassenach ass, Laddie, it's all about to be explained. We'll keep it simple for you. Then, if you're a good boy, I'll buy you breakfast.'

Harry pulls away and is about to retort, Rensa steps between and leads him away.

The main body of the massive Carnegie Space Agency restaurant is partitioned off with movable wall-sections, making a private annexe. The rest of the establishment is teeming with curious people jabbering in multi-lingual conversation. Harry and Rensa are led to the annex, to a long table. Hamish, plus Harry's old team of two years previous, are already seated and eating. Redman sits at the far end of the table with his cronies.

They all stand at Hamish's command: 'Three cheers for Harry, hip, hip.' The group cheer. Hamish gives Harry a damning look, reminding him of his deep continuing resentment. Harry returns the withering look, and then purposely re-directs his eyes to the injured hand. He looks back into the big man's eyes and smiles. Hamish's look blackens as he continues to cheerlead with extreme control, 'Hip, hip.' All respond with resounding hurrahs, then they settle to breakfast once again – Harry electing a full English and a pot of Earl-Grey tea.

To Harry's right is a good-looking woman in her late thirties. It is Helen Cassidy, the one-time colleague of Rose Hawkins, his late lover and mother of his child. Helen eyes Harry for an uncomfortable long time. Harry, feeling her penetrating stare, is obliged to speak.

'Helen, isn't it?'

She is momentarily caught off-guard and answers slightly embarrassed, 'Lo' Harry... long time...'

'Yes, two bloody years is a–' She turns away, cutting him short and leaving *him* now slightly embarrassed.

To Harry's left is Hamish, eating one-handed as best he can. Next to him is Rensa. Directly opposite Helen is a brash good-looking young man in his early twenties, John MacKay. He leans across the table and speaks to Harry:

'Welcome aboard, Harry. I heard all about you from Helen – I'm John.'

Harry, glad for the diversion, smiles, 'How'd' you do?'

'How'd I do what?' says John, slightly puzzled, 'Oh, yeah. Yeah, I do okay, I guess.' He looks across at Helen and leans over and takes her hand in his. She smiles back at him. He continues to Harry, gushing, 'This is Helen... we're going steady.' Helen looks away and continues talking to the woman next to her. John frowns and continues, 'Hey Harry, I'm a silver-surfer too. Second man on Junairo3... you'll be instructing me too, I guess.'

'I'll what?' gasps Harry, choking on his food.

'Yeah, you'll be instructing me, too. —What'd you think of our good ol' boys in grey, have they got balls or have they got balls?'

Harry clears his throat, 'I don't know, John. *Have* they got balls? I never asked.'

'Sure as hell they have, peckers too, haa. See, they need gonads to supply testosterone, can't have no faggot pinko in space, Harry.'

'No, I suppose not, John.'

'You and Helen go way back. She tol' me about you an' Rose an' all. Hey, me and Helen are getting hitched after this.' John gives Helen a quick look. She is still talking. He gives a worried frown, 'I think all she sees in me, Harry, is a ticket-to-ride... SilverCord.' Helen hearing the word turns to him. He winks, she grudgingly smiles and looks away. John continues to Harry, 'Jeesss she's pissed with someone – seems like everybody now-days is pissed with someone. Anyways Harry, good to meet you, and I look forward to learning from you.' He extends his hand. Harry nods 'thanks' and shakes his hand warmly, glad to have at least one friend.

Harry continues eating for a few moments, then abruptly stops. Somebody has touched him in a very personal area. Startled, he looks to his right. Helen is eating, both hands visible. Across the table John is eating, again both hands in view. Harry looks to Hamish, the big man's full attention is devoted to negotiating single-handed a huge plate of ham and eggs, his other hand is innocently nestling in its sling. Harry now looks past Hamish, along the table to Rensa, to find her smiling knowingly back at him. She is leaning on her huge, beautifully tanned left arm, her right arm nowhere to be seen. We see Harry react again as the hand explores a little further. Rensa winks as her right hand now returns to the table, Harry is mortified. Mercifully Major enters and takes his place at the head of the table, and remaining standing he addresses all:

'Okay, lets have some order. Ladies, gentlemen we have an addition to the team. Some of you know Harry Mandyke–'

'Man, drake,' Harry chips in, 'for Christ's sake.'

Major gives a bemused look, then continues. 'Some of you have only heard of him and of his uncle, the late Lord Melrose, author and inventor of the original 'SilverCord Effect'.

Redman, from the far end of the table, interrupts. 'You know, Harry, the original, the one that don't work.'

They all laugh. Harry starts to get annoyed. Major continues.

'For Harry's benefit, I will explain our version of SilverCord. Harry's been away and things have passed him by somewhat.'

'Harry's been in the old slammerroo.' adds Redman, smiling impishly.

Major continues unabashed, 'SilverCord is, as you all know, the name of a technique which enables a mind to leave its body... to levitate. The SilverCord is the link between the mind and the body, the Golden Bowl being the body. And if, while you are levitating, surfing as we call it, you break that cord you can kiss your mind goodbye. Break the Golden Bowl and you kiss your sorry ass goodbye... So say the ancients.'

'That's not strictly true,' says Harry, 'you see– '

'You'd better strictly fucking believe it, Laddie,' interrupts Redman.

'Right on, Woooweee!' chips John, feeling that things are warming up.

More laughter, Major speaks over. 'Okay okay... let's have some order. Now, walking the cord has been achieved before in the past. For thousands of years it has been an excepted phenomenon, only the West had viewed it with scepticism... until the last decade. Then, with the numerous reported Near Death experiences, and the phenomena, Remote Viewing, Shape-shifting, and Bilocation, it was taken seriously and experiments started. But we, the Agency, have the edge... we have Harry.'

'We don't need him,' growls Redman, 'We got it fixed.'

'No, we don't 'got it fixed',' says Major, trying to re-gather his command, 'And yes, we do need him.'

'Well, I'd just as soon leave it to you guys,' says Harry dismissively, not wanting to continue the conversation.

Major frowns and raises his voice above them. 'We digress. Everybody here knows it works. Harry was probably the first of this present company to do so. He didn't relocate to another body, but as we know that part of the process is now elementary. Harry levitated, but with one main difference. He did it over a distance of three-hundred miles.'

A buzz of excited chatter, people lean forward to study Harry as they discuss and speculate.

Redman stands and is visibly angered. 'I don't fucking believe it. If you think I'm giving up my place or John's for that lying, Limey faggot, think again. I've risked my sanity over this project, and I'm going to see it through.'

Major raises his hand for order, 'Look... not Harry nor anybody is taking over as long as I–'

Before Major finishes his sentence, Harry is on his feet. 'The hell do you mean, 'take over'? Up your bloody pipe, matey. And that goes for the lot of you,' he turns his venom on Redman, 'And you can stick your poxy job. I don't want anything to do with it. And don't call me a faggot... those who know me, know different.'

'Yeah, and there's not many of those who know you left alive, you murdering Limey fuck.'

Harry, hysterical with rage, hurls his plate in the direction of Redman and before the missile has landed he has followed it down the table, set on the same course. — Shambles!

As Redman parries the plate Harry is on him, grabbing, clawing and screaming, 'Bastard. Call me murderer again, I'll bloody kill–'

His words are cut off mid-sentence as Redman grabs him, within an inch of a broken forearm, in a tai-chi arm lock. Redman, bleeding from a slight scratch on his hairless head, tightens his hold. Harry screams. —Redman screams, louder. Rensa has picked up the little man by the armpits and applied enormous pressure. Aides converge as the brawl spreads to the other diners.

CHAPTER FOUR

From ugly melee to tranquility: Major's office one hour later. Major seated at his desk. Harry and about a dozen other people including Redman, Helen, and John, stand to attention as if on parade in front of the desk. Hamish and Rensa hover menacingly as sentinels, either side of Major, who now speaks to all but directed mainly at Harry.

'Was that really fucking necessary? Can't you contain your hormonal goddam outbreaks for just long enough for me to speak?'

'Listen…' says Harry, trying to explain.

'No, you fucking listen, Mandyke. I want an apology.'

Redman laughs. Major turns on him. 'Goddam it, that includes the rest of you. You especially, Redman… apologise to Harry, immediately.'

'Okay, sor,ry, Jeezus.' says Redman, amazed at his bad treatment. He grudgingly turns to Harry. 'I'm sorry I called you a fruit. What the hell. Okay?'

'To hell with that,' snarls Harry, far from okay, 'I don't have a problem with gays. I want an apology for the slur about murder. You are the third person who's called me that.' He turns to Major. 'Before we go any further, I demand you explain to this, this assemblage of plankter what happened to Rose. If you don't, I'm out of here.'

Major is about to speak. Hamish butts in:

'What about Rex?'

Major gives Hamish a disappointed look, 'Goddam it, that was two years ago. That will do.'

'No, it won't 'do'... I need to know about Rex?'

Major ignores Hamish and continues to Harry, 'Okay, the bollocking, I believe that's what you Limies call it, is over. All of you, take a seat where you can.' They all sit in the chairs that line the office. 'Now,' he turns to Harry, 'they already know about what happened, Hal, but it's just hard for some of us to accept. Mainly because there's no proof, no evidential confirmation... nothing tangible... it was all destroyed. Give them a break and I'm sure you'll get the same.' He looks at the group one at a time, then continues. 'What in hell do you suppose happened on Junairo2, for Christ's sake? You think we jettisoned three of our own people in space for fucking kicks? You saw the data. Goddam it, what more do you need? You saw those malformed bodies. What did you think did that to them, warp fucking drive?' He looks directly at Harry. 'We all lost friends... Harry lost more than all of us.' Rensa glares at Major. He acknowledges and corrects, 'Sorry, more than most of us. Now, you all know about the British 1950's manned shot around Mars.'

'You mean the one the Brits traded jet-engine technology with Russia, for their captured Nazi atomic motor,' says Redman in the nastiest possible way, 'so the fucking Commies could put the engines in the Mig 15 and kill good ol' American airmen in Korea–'

'That's not true.' interrupts Harry.

'The hell it is. They traded it for a captured Nazi rocket motor. They used it in their fucked-up, cockamamie high-polluting space shot. —That the one you mean, Major?'

Rensa moves to Redman and glares at him, he laughs in her face. She keeps staring until he has to look away. She returns to her position and Major continues.

'Yes... that's the one. An entity got into the ship and

killed the crew. And you know about Junairo1, the same scenario: Of the three crewmembers, two, the men, were killed and the flesh was used... and with it the woman occupant, cloned, duplicated herself, whatever,' he looks to Harry.

Harry answers the look. 'Yes, that is so. There's an equation for this process. It's as follows: the entity uses two bodies, one to copy, and half of the other to use as fuel. It creates a living crucible.'

Redman, disinterested, is talking to Helen and totally ignoring Harry. Major turns on him angrily:

'This concerns you too, Redman.'

Redman stops and looks up at Major. 'Sorry, Major,' he turns to Harry, 'Sorry Harry, continue, continue, I'm all ears.' He pulls out his ears, like Mr. Spock, and smiles.

Major calms and takes up the explanation once again. 'As Harry says, it uses just half the second body, that's why we found so many torsos, that's why Rose Hawkins reproduced two of herself... if you work it out there wasn't enough to make herself and one of the men.'

After a few moments of grudging attention, Redman returns to talking and laughing with Helen.

Hamish gives him a threatening look then moves and joins Major, he speaks covertly: 'Do you want me to intervene?'

Major shakes his head and continues to all. 'So, the entity doesn't think for itself... it's catalytic. It just uses the highest intellect available. It kills to survive... tongues come out of its mouth and stomach and it eats people – Redman! Goddam it, will you listen?'

Redman stops, smiles and paddles his hands as to say, *take it easy, take it easy.* 'Sorry Major, this is elementary to me, but I'm sure the great Harry Mandrake would like a refresher.'

'Listen to me, you impudent little man, I don't give a fuck if Harry's put your nose out of joint. I'm not a scientist, I'm in the business a making a buck. If you want technology go punch a computer,' Redman shrugs and smiles. Major continues, 'If it's a toss-up between flesh and money... well, you know what they say, 'the flesh is weak, but diamonds are forever'. That's my philosophy, take it or leave it. Just remember, if it wasn't for a buck, we'd still be waiting to get to the moon. It's all to do with money and your attitude, Redman, 'he glares at Redman. 'is costing me FUCKING MONEY.'

'Sorry Major,' says Redman hiding his face, but everybody including Major can see he is rocking with laughter.

Major composes himself and continues. 'Every word is true. It eats people to survive, and, like the Medusa it turns people into ash. You all know this, so give Harry some space. We need him desperately.'

Redman, now paying full attention is sitting bolt upright with arms folded in a parody of an attentive schoolboy. Major gives him a withering look. He then turns to Hamish, leans toward him and speaks covertly. 'If you can't cut it, Son – you know, Rex an' all – you'd better say so, I don't want no vendettas, I've got other things you could be doing. It's up to you.'

Hamish smiles, 'I'll be okay, I can handle it... just having a little fun.' He turns from Major and walks to where Harry is sitting.

Harry looks up at him suspiciously, 'Got something on your mind, old boy?'

'A truce, Hal? What say *old boy?*' Hamish takes his hand from the sling and offers it to Harry to shake.

Harry gently shakes it. 'Why not.'

'No hard feelings?'

Harry nods agreement. 'No hard feelings?'

Relieved, Major claps his hands for attention. 'Right, good. Let's get to it. Yes, we levitate; we can climb the silver cord, and, through a technique we've developed from your uncle's theory Harry, we can transpose a human mind to a latent recipient synthetic body. A man in proxy.'

'You can actually do it?' says Harry.

'Yes, you've seen it work. Our problem however is distance. Your uncle theorized that distance would not exist in this concept, nor would time... the next room or a light-year, or infinity. That's why we need you Hal. We've hit a quarter-mile barrier.

'And I hold that record,' chips Redman.

Major nods verification, and continues. 'When we try to go further... well, I'll let the record holder tell you. Red:'

Redman stands and takes an impudent bow. 'So, when I said I nearly lost my mind, I wasn't kidding... exactly that. I got lost. I never felt so utterly alone in my life. They had to bring my body back into range.'

'What did you find out there?' asks Harry, politely.'

'Out there–'

'Wooowhee,' interrupts John, gushing his answer, 'I call it the fockin' Badlands.'

Redman, not liking being interrupted, ignores John's comment. He answers Harry curtly, 'God knows what's out there, Laddie. I certainly don't.'

John proffers his opinion again, 'Badlands.'

Redman gives him a viper look then continues, 'Blackness... Badlands if you must, the stuff that dreams are made of, bad fucking dreams... nightmares. But what's more to the point, Harry, is how did you manage over three hundred miles, to a moving train I understand?'

'Simple,' says Harry as if the answer is elementary, 'I had the British Railways time-table.' They all look astounded. We see the first ray of humour returning to Harry's face.

Redman is not so amused, 'What the hell... ?'

'Joke,' says Harry, now openly smiling. 'How the hell do I know, it was more than twenty years ago? I'll need to re-think and to re-study the file. I need to re-create the circumstances of that trip... I was only sixteen. —Now, when am I going to see my son?'

Major, ignoring Harry's pertinent question, continues to all, 'Okay, let's break on that.' He turns to Harry. 'I'll get you a copy of the file.'

They start to walk out. Harry steps in front of Major, obstructing his way. 'No, not a copy. I must have the original file... please. I just want to have it in my hands, it'll help me remember. And now the question of my son?'

'Okay, the original. But your son must wait, you've got work to do. Oh yeah, and you've got your old office back... all your stuff is there, all your faggot clothes an' all.' He stops for a moment and considers. 'Sorry, Hal, seeing your son now would only complicate things.'

'Okay. When? I need a firm date. I don't move from here till I get it. I mean it, Major.' Harry refuses to move out of Major's way. Major studies him eye-ball-to-eyeball... uncomfortable seconds tick past. Major clasps his hands around Harry's shoulders and tries to move him on. Harry stands his ground, unflinching. 'I mean it, Major.'

'Don't worry, Harry, you'll see your son. Give it a week.'

Harry adjusts his watch, 'Right, a week, exactly a week

from now... I make it thirteen hundred hours, near as damn it.'

Major nods in insincere agreement, Harry concedes and they walk out of the office together. They catch up with the others and together enter the restaurant. John approaches Harry. He speaks covertly:

'Excuse me Harry, I'd like a word. I've no hard feelings about J3, I just want the trip. I'd understand if you did want in... SilverCord being your baby an' all.'

'J3?' says Harry, puzzled.

'Junairo3. Thing is, Harry, I want Helen out. This shit is dangerous. I got lost out there... the fockin' bogeyman is out there, man... I mean it. She's determined to surf, an' it's no place for a lady. Will you try for me?'

'Certainly, John, I'll certainly try.' Harry smiles, flattered by John's misguided opinion of his authority. John walks on.

Redman now approaches. 'I'll no be sitting with you at lunch, Laddie,' says he, leaning into Harry's space, 'I don't want your lunch in my lap as well as your breakfast.'

Rensa now sticks her face into his. 'Don't worry, you impudent pink-eyed dwarf, I'm looking after Harry from now on. He's in my charge, and that's official.' She moves away taking Harry's hand and leads him off like a mother with her fledgling.

Redman laughs and calls out after them, 'Watch your wee arse, Laddie.'

CHAPTER FIVE

Redman's impudent face now stares passive and inert. He is cadaverous still, lying in a glass coffer, secured in the cargo hold of SBS Orion, now flying on the edge of space. Harry's face imposes over Redman's as a flimsy, ethereal second image. A second SBS, carrying a donor synthetic, is shepherding Redman's craft. It fires its rocket boosters that will take it three hundred miles into the stratosphere.

'Come on, old sport, you can do it,' Redman's closed eyes react, darting fitfully under his lids, in answer to Harry's disembodied voice, 'It's out there... believe. You're thinking too much. Don't rely on the senses. Let the id take over. It's just one step, a yard, a mile... a million miles... just one step.' Harry's face degrades and the face of a grey synthetic is now a second image. 'You've got to will it to be there, old man, you have to believe... you are as Christ walking on water – believe.'

Redman's face superimposes over the synthetic face. The two images merge and dissolve away into the mist of the Badlands, the hellish nightmare-land of the disembodied mind. Redman stumbles through a horror dreamscape, arms extended, clumsily dragging himself over unseen obstacles. His eyes have no pupils, just pale sightless irises set in wide-open whites. In this horror netherworld, he is not alone. There are other things lurking in a swirling mist. Dark incomplete things, incarnations of human figures, tangible essences of real people, some staggering some crawling, some accidentally converging into each other and holding on in despair. Some of these spectres appear to have solid form and feature, none seem able to see or communicate. They appear lost and in absolute terror – in Limbo.

Harry's voice calls through the void. The mist parts and a silver ribbon of light stretches out in front of Redman. He turns unseeing, into the direction of Harry's voice calling from the silver cord of light, all the time willing, guiding him onward. Redman's terrified face is now layered with the grey face of the synthetic. At the end of the silver beam of light the SBS appears. The synthetic's eyes open, superimposed over Redman's sightless sockets. The grey face and SBS slowly dissolve away, leaving Redman lying in his coffer, attached to wires and monitors. Harry stands over him.

The lid of the coffer opens and Redman sits up. Gradually he gathers his wits. He slaps at Harry's open hand and yells, impudently mimicking Harry, 'Bloody bingo.'

'You did it, old luv,' says Harry, jubilantly, 'three hundred miles, and much more – you're ready for infinity.'

Harry now sits snug in his old Carnegie Agency office. He is working at his huge roll-top desk – the same one he'd used just over two years ago. Strewn out before him is the pink file marked 'SilverCord.' Every corner of the office is packed as before with electronic IT peripheral, ancient and modern, plus his trusty Krups coffee-maker and plastic cup dispenser just as he'd left it an era ago. He stares, mesmerized by the battered and faded cerise dossier. A buzzing invades the tranquillity. Harry flicks the intercom. A female voice gives forth:

'Mr. Mandrake, a Miss Rensa Lansavich to see you. I've sent her up.'

Harry answers, slightly annoyed, 'Okay, but next time bloodywell ask, first.' He clicks off the intercom before the woman can answer. With a remote-control he switches on the coffee machine then quickly tidies the office. He opens one of the big metal cabinets containing his stored clothes: each item individually packaged in a hermetically sealed cover. He takes out a Harris Tweed jacket and removes it from the plastic. As he is about to put the jacket on he notices a yellow cleaner's ticket still pinned to the collar. He fumbles to remove it, stops and considers. He leaves it where it is and slips the jacket on, the label clearly visible hanging from the back, then undoes a fly button, and then quickly sits back at his desk.

Rensa enters. She looks absolutely devastating dressed in a strapless off-the-shoulder powder blue dress, long gloves, slick-backed hair and stiletto heels etcetera, a paradox of delicate provocative sophistication. She stands towering a huge six foot six plus in the doorway. Lit from behind she rotates as she speaks, oozing confidence as to the expected answer to her rhetorical question, 'What do you think, Harry?'

Through Harry's transparent leer, she knows exactly what he thinks. 'Very nice. Very, very nice... and you'd be shocked if I told you what I really think.'

'Shocked... don't, be, ridiculous. How did you get on with Pinky and Perky?'

Harry frowns, then falls in. 'Oh, you mean John and Redman – very good. Yes, we made good progress. What do you think of John and Helen getting it together?'

'Why not? She's a bit long-in-the-tooth for him, but

good luck to them... if you can find a bit of, shall we say love in this shithouse world, Harry, why not? I always thought she was AC DC...' she paddles her hand. 'Anyway, why not?'

'Yes, I remember Rose saying she'd made a pass at her once.'

'Hey, it's about time you made a pass at me, or do we talk about your old flames all night?' She exaggerates her figure into a voluptuous S. Harry's mouth sags open – his two years of celibacy welling to bursting point. She puts her finger under his chin and closes his mouth chomping his teeth together.

'Now, turn off your coffee maker. I've come to take my boy out on the town.' She leans over his desktop, looks down at him, noticing the label on his jacket she leans further over him, forcing his head into her cleavage.

'What are you doing?'

She removes the label. 'Taking you out of mothballs by the look of it.' She drops the label on his desk. Harry acts surprised.

'God, that's been there all day. I've been away a long time, Rensa. I've forgotten how to act with a lady.'

'Rena, please. I'm not Russian, for Christ's sake; I was born in Idaho. My father was a Ukraine migrant. Major likes his little joke.'

'So, Rena, I'm in your charge... day and night?' He stands up and into her arms... she towers over him. He runs his hands up from her waist to her bare armpits and pulls her to him. She has to bend slightly downward for him to kiss her. Harry responds by moving his hand to cup one of her huge breasts, schoolboy to schoolgirl fashion. Rena adjusts the direction of his hand back to her waist as she pulls away from the embrace.

'Not here. If we do make love, darling Harry, it won't be on your office carpet.'

'I wasn't trying to–'

'Shut your mouth and do up your fly, Mister. We're going to hit the town.'

'Suppose I try to escape?'

She kisses him again, a long open-mouthed, smothering kiss, bending him backwards, virtually taking the male pose. She speaks as she pulls away. 'You won't escape, Harry.'

Harry, slightly embarrassed, tries to hide the bulge in his pants as he dose up his fly. He closes the pink file and locks it away in a metal cabinet. Together they leave the office.

'What's out there, Harry?' says Rena, as they approach the open lift, 'I need to know…' she mimics John's voice, '… in'a fockin' Badlands?'

Harry shrugs, 'There's nightmares out there, lost pieces of real people… and creatures dreamed into being by… who knows?'

'Help me out here, Harry, hazard a guess. I really do need to know.'

'Dissociated schizoid personalities, the comatose sick, the near and newly-dead, and,' he gives a little laugh, 'SilverCord surfers, lost in Limbo… the place in some Christian beliefs for un-baptised children. Oh, and those who died before the coming of Christ… and… God alone knows what else.' But they're there… in purgatory, for all I know.'

'Damn it, Harry,' says Rena as they enter the lift and the doors start to close, 'you've left your coffee-maker on.'

'Don't worry, it'll turn itself off after–'

The lift doors close... buckle and implode in surreal motion. Harry's office door glides down the corridor, shards of toughened glass tumble gracefully beside. The accompanying protracted drone reverts to an ear-splitting crack of an explosion – chaos. Then, like a hammer blow, silence.

After a few moments, one of the distorted lift doors slowly opens, the other refusing. Rena emerges with difficulty, unshaken. After a few more moments Harry emerges, very shaken. He slowly staggers to his office and joins Rena, who is already attending the small fire around his desk with an extinguisher.

'See,' says she, blasé, 'I told you to turn your coffee-maker off. Now we're going to be late for the Opera.'

'A bomb,' weeps Harry. 'A bloody bomb! Some bastard is trying to kill me.'

'Or me,' says Rena, 'or destroy the SilverCord file. The file's okay, by the way.'

Harry not hearing, 'Some bastard is trying to kill me. Oh, my, god.'

The fat lady sings... the aria from Aida. Harry is sitting next to Rena in the luxuriance and safety of the opera house. He is now calm, lost in the sweet music. In the delirium of the atmosphere he begins to doze, entering troubled dream-sleep. In his dream he is slowly gyrating, dancing in time with the music. He has a woman partner. Her flowing hair momentarily obscures her features... Now he sees her face, it is Rose Hawkins, his deceased lover and mother of his son. She smiles at him and they kiss, a long passionate kiss. The music is beautiful, exquisite. But as Rose tips her head back there is blood on her neck, a trickle just below her ear. It runs down around her neck into her beautiful necklace, the cluster of pearls and coral-coloured pellets. Harry gives a growling moan as he desperately tries to break from the dream and back to the safety of reality. The necklace noiselessly explodes, throwing out a convulsion of blood and tissue. Harry screams out, breaking from his dream:

'Naaaa naa... no, Christ.... NOOO!'

The dream fades. His outburst has disrupted the performance and the theatre is now deathly quiet. After a few moments, people start to mumble. Guardedly the players on the stage desperately await instruction. Ushers descend and Rena helps Harry out of his seat. As they exit the conductor calls the orchestra to order and the music starts again.

Later that night, in the sanctuary of her apartment, Rena and Harry embrace on a massive black-sheeted bed. Rena is naked, tanned and contrasting beautifully with the blue moonlight, her Amazonian figure engulfing Harry's puny, sun-starved body... a giant to a pygmy. A delight of sensual foreplay: Harry's repertoire of well-practiced maneuvers, having lost nothing in their enforced idleness, are employ to the full. Climaxing, at length, in the inevitable simultaneous consummation of newfound love.

They lie awhile in silence.

'I'm dog-tired, but I daren't sleep,' says Harry at length, his eyes wide open and staring into the moonlight. 'I'm too bloody scared to close my eyes, or even to blink.'

She hugs him and whispers in his ear, 'Don't worry, Harry, Rena's got you. I'll make you sleepy.'

After a brief encore of lovemaking, they both yield to the cloak of Morpheus as the blue moonlight concedes total darkness.

The clouds move on and blue light is back, brighter than before, illuminating Harry as he sleeps. He reacts with a sleepy smile to a beneath-the-covers caress – or so he imagines. The black silk sheet dampens with glistering blood-slime, fluid oozing through the fine weave preceding the blood-coloured tentacles. First one, then two, and now a dozen. The bed is awash with digestive slime and alive with a hundred scarlet, searching tongues. Harry sleeps on as the tentacles seek out and penetrate his ears, eyes, and nose. He wakes with a coughing, choking fit. Hardly able to find his voice he gives a child-like wail. Across the blooded sheet Rena, pink, naked and beautiful in spite of the thick coating of blood and slime, casually enquires, 'What's the matter, are my hands cold or did I grab something too hard with my big, clumsy mitts?'

As Harry again tries to speak, one of the tentacles enters his mouth. He closes his teeth on it, slowly biting it in half. Blood and slime drool from his lips. He takes the bitten tentacle-tip out of his mouth with finger and thumb as if it was a piece of macaroni and, whilst wiping away the bloody slime, speaks as the recurring nightmare starts to dissolve away.

'Bad dream, old luv. It's okay, some I can handle. And no, you naughty little girl…' he gingerly opens an eye, the slime and tentacles evaporating to the conscious world, '… your hands are just the right size.'

Rena, pink, naked and beautiful, sans blood, sans slim smiles, 'Why, thank you kindly, Sir.'

'Think nothing of it.'

'So, the bogeyman is gone. What now?'

Harry also smiles, *'Behold, as may unworthiness define, a little touch of Harry in the night.'*

'Shakespeare,' chips Rena, 'Henry… four, five,

something, I think – And it's nearly morning.'

'Henry the Fifth – very good. Yes, nearly morning, so we'd better get a wiggle on.'

'Wiggle? …Oh, Harryyyy!'

They fall to lovemaking, which carries on from the bed into the bathroom and on into the shower. Kissing and frolicking in the steaming water, Harry soaping and caressing her at the same time.

After the shower, Harry dries her off, and powders her all over with a big powder-puff. Giggling, Rena takes a handful of the powder from the box and smothers it into his groin. Harry gives a little choking yelp and they start their naked frolic again. They toss and tumble, dust and spray, squirting and splashing each other with the toiletry product delights of the dressing room. Then, powdered creamed and gelled, they shower again.

Playtime ended Rena fusses over herself in the dressing room mirror. Harry, now in his robe, proceeds to unpack his one suitcase.

'Okay, I'll stay, just for the time being mind,' says he candidly over his shoulder, 'but I must have an apartment of my own. They can bloody well afford it – I'll make the buggers pay.' He unpacks his pair of enormous dental syringes, still held in his custom-made leather shoulder holster. He looks at them and considers. His thoughts flash back to pregnant Rosette, she ripping her stomach open, slime and tentacles bursting out at him and he rendering her unconscious with an electric stun device, and finally pumping syringe after syringe of novocaine into her.

He shakes away the horrific thoughts and calls out to Rena, 'You know the score… about what happened to Rosette? You know I killed her… with these… filled with novocaine? I killed Rose, too with–'

'With an exploding necklace.' Rena finishes the sentence as she walks from the bathroom, now fully dressed. She sits next to him as he continues inspecting his syringes. She is quiet for a moment as she studies him. 'I know it all, Hal.'

'Have you seen my–'

'Your son? Yes, I've met him,' she gives a mocking smile, 'He's very nice. Not a bit like you, he's got character.'

'Good,' says Harry, nonplussed, 'then he won't be bullied by pushy bloody women, will he?'

Rena laughs. 'Anyway, you won't need those things anymore. Better let me have them.'

Harry turns away so she can't take them, 'You think so? I'm not so sure. It's still out there. I'm not going to end up a burned up pile of slime.'

'Nothing is out there, Harry.'

'Yes, there bloodywell is. And it's trying to kill me. I'm not taking any chances, whatever you say.'

'What I meant was, Mr. Shitty, we've developed a spiked bullet. It's got a hollow nose filled with concentrate novocaine and mercury.'

'Oh, really' says Harry, hardly interested.

'Yes, really. 'Blue-nose' we call them. We couldn't improve on novocaine... we didn't have anything to test it on. We just improved on its delivery power.'

'Why the mercury?'

'It's a hollow round... when a blue-nose hits its target the inertia hurls the mercury and bursts the bullet casing... Pow.'

Harry makes a gun with his finger and fires back, 'Pow, pow, pow.' fanning the imaginary weapon cowboy-style. She now makes a hand like a gun and pokes the pointed finger under Harry's armpit with some little force and screws up her nose.

'Ow!' He recoils with instinct reaction.

'Anyway, that's the theory.'

'Great, so if you miss the bad guy and hit the good guy, at least it doesn't hurt... gives friendly fire a totally new meaning. You could probably patent them... 'Make war fun with Blue-Nose bullets, the friendly way to kill your enemy'... I'll stick with these if you don't mind.'

She gives him a tiresome look, 'There's nothing out there Harry. It's been two years. We'd have known by now.'

'Oh, so 'nothing' blew my office to bits.'

'The bomb was something else. Russia is back in the equation... with a vengeance. They want their Mars portfolio back, or a bigger share.'

'Then give it bloody back, or give them a bigger share.'

'Don't, be, ridiculous. Anyway, they are accusing us of international looting. If we give it back it's as good as admitting it. We bought that stuff... millions of dollars... zillions.'

'If the material was paid for, what's their problem?'

'Ex,actly. Anyway it is a problem, and it is a fact. They are taking us to the International Court for its return, and with our pussy government, they'll win.

'So?'

'So, that's why we need that manned landing. If we don't set foot pretty soon, that's it... we've had it.'

'Oh dear – what a shame – never mind,' says Harry thumbing his nose.

'There's a lot of people who'd like to see us fall on our asses, Hal... the government for one.'

'Me, for another.'

'You don't mean that, Harry... your uncle's work will come to nothing. If we screw up, the contract is gone and the US Government will go-it-alone and save themselves a fortune.'

'But they're still out there, Rena.'

'That's why we've brought SilverCord forward. We didn't want to send another human crew. The shield wasn't strong enough. It didn't stop it, but it did affect it.'

Harry shrugs, 'So what?'

'So What?' She turns on him angrily, and for the first time losing her cool. 'So men where burned and fused together, Harry. I demanded to see the footage. It was horrible, HORRIBLE.' She puts her hands to her face.

Harry stands up and holds her. She buries her head in his shoulder and weeps.

'Hey, what's this?'

'My *father* was on that ship – Your bloody shield didn't work.'

'Dear God, I'm sorry. I didn't know that. Nobody tells me anything. I told them it wouldn't work. It needed development... I told them. It must have been hell for you to watch.'

She wipes her eyes, 'Sorry Hal. I'm not angry with you. It's like Major said, 'we are all expendable'. What must you think of me, I'm acting like some stupid kid?'

'Not at all, I know what it's like.'

'Both your folks were killed when you were a child... I read your file. It must have been hell for you, too.'

'I didn't know them that well. I knew my uncle better. I spent most of my time with him... holidays and that.'

'How come?'

'Don't know really, we just got on. That's how I got to

know of SilverCord. That's what he was working on that summer. He taught me everything. I think he was schooling me for this. He knew we'd go back, and that it would still be there.'

'You loved him, Hal?'

'Yes, like a father. Hey, he's not dead you know, not definitely, just bloody missing. God I could do with him here now.'

'At least I know – closure.'

'Yes, losing my parents was hell. That's when I used SilverCord. They let me come home for the funeral. I decided I wasn't going back. That school all but robbed me of my parents... Anyway, that's the way I saw it.'

'But they did send you back?'

'Yes. I couldn't stop them. But I vowed they wouldn't have my mind. God knows what I was about. They put me on a train – sixteen years old and just buried my parents. That's the way it was, then—'stiff upper lip, and soldier on'—Well, to hell with that. I wasn't having it.'

'How sad. They should have let you stay a while. So, what did you do?'

'I just split, climbed the cord and 'surfed' as you call it. I went to my parents' wake. I saw my aunts, uncles, and cousins.' Harry closes his eyes for a moment and remembers: A country mansion at Melrose, the Scottish Borders, a wake. Among the many guests are the ethereal spectres of his parents. He opens his eyes and continues. 'Then I saw my parents, just sitting among the guests.'

'You saw your parents?' says Rena, shocked.

'Yes, and they saw me.'

'Jeesus H Christ.'

'Don't blaspheme.'

'Sorry. Carry on, your father...'

'My father just gestured with his hand 'go back,' as if I'd just stepped into his study. I was bid 'make yourself scarce, and get back to school' – my mother smiled, then together they were gone. So I did as I was told, I retraced my steps back to school.' He closes his eyes again and he sees his parents walking off into a flare of light.

Rena leans to him, kisses his cheek and the vision is gone. He carries on unpacking his bag. He stops and takes out a little platinum locket on a chain, and hands it to her.

'I want you to have this. I gave it to Rose a long time ago. It's been in the family–' he considers, then continues, 'Skip that. I just want you to have it.'

'Thank you, Harry, it's beautiful; it's not going to explode is it?' Harry gives a frown. Rena is delighted. 'I'll take it on loan. When we split, you get it back... it's all you've got of Rose.'

'Not all... I've got Barney,' says Harry, doing up the clasp. 'When this thing is done, Rena, I want some time to get to know you; you suit me, we fit nicely together.'

'Come on, lover-boy, we got to go, before you knock me off my feet with your Limey flattery.'

'Go where?'

'Back. We must go back. There's a pre-brief meeting with security. They're crapping themselves about the bomb.'

'Bloody hell. Must we?'

'Yep. They want us all to be there. Major, Hamish, Redman, Helen and John, the whole SilverCord team.' She gives a mocking smile. 'You can bring your syringes and give them all a jab if you think it will prove anything.'

'Damn good idea. Do you think they'll let me do it?'

'Fuck Hal, I was joking.'

'Please, pleeease don't swear, Rena. I hate to hear

women swear.'

Rena gives him a long questioning look. 'Don't swear, don't blaspheme, don't... *knickers* to your don'ts. How does that grab you?'

Harry gives an impish smirk, 'I say, I like it... very English. I like it a lot.' He hugs her, obviously turned on by the word. He tries to lift her towards the bed.

'Put me down you dirty old man', says Rena, through her laughter, 'you'll hurt yourself. Christ, only a Brit could be turned on by a stupid word like 'knickers' for pants – and, what's the other one, 'knockers' for breasts.'

'Just a quickie.' In his excitement, he stumbles and has to let her down.

'Come on, Harry, playtime's over, we've got two meetings, then we leave for EarthLab, at noon.'

CHAPTER SIX

Through the velveteen indigo of space, EarthlabOne emerges from a single dot of light: a shapeless myriad of decks, causeways and gantries. Tethered along side is the Junairo3 spacecraft, and about ten nautical miles away lays another craft under construction, the Argos.

An SBS approaches and fires its stern gas-jets and closes on the great confusion of metal. It slings and secures its umbilical lanyard and slowly manoeuvres itself to the space station, then docks and enters the airlock. Once inside the transit-station the bay doors close and the SBS airlock opens. Through the fine mist, Major, Hamish, Redman, Helen and John, Rena and Harry disembark. Acting Commander McQueen, a good-looking woman dressed in civilian executive-type clothing and sporting a grim no-nonsense persona, meets them. She greets them with a curt, 'Major. Everything is ready.'

Major smiles and nods, 'Thank you. Nice to see you looking well.'

McQueen doesn't reciprocate. 'I'll about my business. You have the freedom of the ship. If you need anything just call – *don't* press any buttons.' She turns and walks off without the hint of a smile. In the quarter Earth gravity, Major and the group amble into the brightly coloured network of corridors.

In the large circular command room, Major takes position at a fixed lectern as the rest of the group sit. He studies them for a moment, then begins.

'Okay, to reiterate, we've been through this before, so we go through it again: Mind transposition is to be achieved onboard the shuttle exactly as it will be in the real thing, that's in two weeks time. For that trip – the last of the Junairo starship missions – Junairo3 will be unmanned, just two latent synthetics waiting in the lander. It will be under maximum power all the way, reaching one-third light-speed and then some.' He is momentarily interrupted with shocked gasps and murmurs. 'I know, I know,' he continues, 'but priorities demand ass-on-planet. It'll be synthetic ass, but ass nevertheless... She'll use her entire fuel capacity on a one-way mad dash.'

John stands and yells, 'WoooWee!'

Major raises his hand for order. John sits and Major continues. 'We effectively park in orbit around Mars' moon, Phobos, and wait for the new starship, Argos. That's just finishing construction here at EarthlabOne, as you can see through the aft porthole. On which will be a manned human crew and two synthetics. In residence will be Redman and...' he gives a little smile, 'Don't worry, Harry, you're not one of them... The second-man will be John–'

'Right on. WoooWee,' yelps John.

Major raises his hand again, John calms. Major continues. 'And you, Helen, will also be aboard as standby. The two ships will dock, Red and John will walk the cord and locate with the synthetics – the landing will be effected with Junairo3's tried and tested landing module. Mission accomplished, they will relocate to the Argos. Junairo3 will stay permanently in Mars' orbit as observation platform. So... today's dress-rehearsal will be conducted from Junairo3, on which will be four non-combatant EarthLab crew plus Redman, and Harry as stand-by for this exercise only. Redman, obviously, will be the main player. We have four such rehearsals scheduled, so everyone gets a turn, Harry will be present in each, merely for guidance. All this you know. Any doubts, check your flight notes. For today's little jaunt, the countdown starts as I speak. Any questions?'

Harry sticks his hand up. 'Yes. I–'

'It's okay Hal... just as stand-by... only in an advisory capacity.'

'Yes but–'

'Contrary to common belief, Junairo3 was not a complete failure. It successfully soft-landed its cargo of equipment and supplies. Also, as experimental expedient, it had sixteen types of living animal tissue on board, including two live huskies and a goddam wolf. Of whom none was in any way affected. So we can conclude that whatever it is that's out there, it attacks only human flesh... the synthetics, being made of animal tissue are also impenetrable.'

'Yes, but what–'

'I'm coming to you, Hal. Now, if we can just confirm that we've overcome the distance barrier—at least three hundred miles—that will be enough for today's exercise. Once Redman has climbed the cord and has located with his donor synthetic, he will take Junairo3's lander and effect the landing – for this exercise, it will be on EarthlabOne.' He stops and gasps a breath.

'Yes, but what about–'

'Nearly there, Hal. Then he will simulate a docking with Argos. After which, he will climb the cord and relocate with his dormant body on EarthlabOne... end of exercise. Are there any questions?'

'Yes. What about me?'

'Good, excellent.' Says Major, still ignoring Harry. 'Any queries, consult your briefing papers. Now today Harry will be in the top tier of coffers.' Turning to Harry. 'There will be a second synthetic next to you, Harry, just for co-ordination. It won't be used, but you'll be in SilverCord too, ready... that's why we've given you a complete rundown on SBS procedure... ready for anything. Ready to walk out there and show the way back if, God forbid, it becomes necessary.'

'Hey, what is this?' yells Harry. 'I said I would cover in an emergency, not as a matter of due cause. You're a bloody liberty-taker.'

Redman stands and shouts above Harry, 'Fuck him. I don't need him. If he's not up for it send him back to the fruit-farm. Stop pussying around and let's burn rocket fuel.'

Major rolls his eyes in exasperation, 'Okay okay, but Harry's right... only in an emergency, and that's all.'

'And I told you, I don't need him.'

'He's done a good job on you, Red. In just twenty-four hours of Harry's influence you've virtually cracked the distance problem. Nothing is going to go wrong.'

'Dear God, I don't need this,' says Harry getting up, as if to leave. 'I want to see my son, or the lot of you can go stuff yourselves.'

'Going somewhere?' says Major, reminding Harry that he is in close Earth orbit. 'Listen to me, Hal, SilverCord is as important a discovery as fusion, more so... it's like splitting the Atom all over again. If you help us, afterwards, you and your uncle's name will take the credit. Just consider the medical benefits: schizophrenia, paraplegia, long-term chemotherapy, organ re-growth, virtual travel, and a zillion other uses.'

'I want to see my son,' yells Harry, now on the brink of fury.

'It's potential is mind-blowing, Harry,' continues Major, ignoring the outburst, 'When we are done you can write a paper and put your name to it 'The Mandyke Effect' you'll be able to name your price. A Nobel Prize, at least.'

'Man,*drake*, for God's sake.'

'Whatever – I know you won't pass this up, Harry… all those years of ridicule, all those senile old farts professors at Oxford – '

'Eton.' says Harry, starting to take interest, 'It should have been Rugby, but some flashy ancestor of mine queered-the-pitch.'

'Queered the pitch. What the hell does that mean?'

'He got sent down, old man. So… Eton.'

'Wherever – You won't wait to shove your paper up there asses. Hal, this is your chance. SilverCord is the passport to the stars. Just imagine, these synthetics will be the new pioneers, the new Founding Fathers. They'll show the grit and take risks while our bodies are safely tucked on EarthLab, but not for this dress rehearsal… they'll be on Junairo3, EarthLab for the real thing. It's inspired, Harry, and it's all yours.'

Harry is silent. He considers for a moment then sits. By the look on his face it is obvious he's hooked. 'Well I'm bloodywell taking these with me, so there.' He opens his jacket, showing his full syringes.

Major gives a little shrug of triumph then continues to all. 'Okay, Harry will be in the second pair of coffers, in silvercord, just waiting with a synthetic in place, as precaution only, as I say just waiting, just as contingent.' He turns and directs his words to Harry, 'Nothing can go wrong Hal, nothing. As I say, for contingency.'

'You don't listen, Major,' yaps Redman, 'I told you I don't need no goddam pussy contingent. If he wants to boogie, fine. If not, let him go 'stuff *himself*', whatever that means.'

Major sighs, 'Okay, let's do it.' He stands and they all start to exit.

Redman hangs back. When he and Major are alone he speaks. 'You wanted a word in private, Major.'

Major studies the little man. 'Close the door,' says he with a hint of menace in his voice. 'Yes, a word with you, in private. I want that... fruit fuck out. He's a goddam liability.'

Redman is relieved and encouraged. 'Now you're talking. And that crap about him taking the credit, you can't be serious... have you forgotten Rex?'

'No, I haven't.'

'Good, because either that murdering cocksucker is out or I'm out.'

Major studies Redman again. They stand staring at each other. Redman calms slightly and then starts to look uncomfortable. Major breaks the silence.

'Don't ever threaten me, Red.'

'That wasn't a threat, but–'

'Let me give you some advice,' he leans further into Redman's face, 'When you go looking for a shit-fight, don't pick it with a skunk. Now, I'll tell you what you're going to do...' Major's face resorts to a mask of unbridled treachery.

Redman and Harry lay prostrate inside Junairo3, one in each of the two linked pairs of coffers. In the adjoining coffer, lying next to each of them is an inert synthetic cadaver.

The four non-combatant Junairo crewmen harness themselves prior to reducing gravity. They call out the checklist:

'Make ready to Purge.'
'Slinging the jamb.'
'Confirmed, jamb slung.'
'Reducing to zero gravity.'
'Zero gravity, confirmative.'
'Vacuum in centrifuge housing confirmed.'
'Starting centrifuge.'
'Centrifuge confirmative: Running.'

The pair of coffers begins to spin on the circular turntable. Above these, the duplicate set of coffers spin in opposition, Harry in one, a synthetic in the other. The spin increases, unhindered by air or gravity, to such speed the coffers become a continuous blur. The deck-lights dim and a series of strobe lights fall across the spinning turntable. One of the crewmen barks out orders:

'Reaching maximum revolution... SilverCord within twenty-five secon–' the man is dead before the word is finished. Shrapnel from a small explosion in the cabin wall rips through flesh and machine. One of the bottom two coffers shatter, throwing out a shower of crystal fragments as it continues to spin. The explosion has punctured two of the three skins of the Junairo3 starship. Of the four crewmen, two are already dead, the other two badly wounded. They scream out incoherent commands. The cabin fills with smoke as a lone vague figure fumbles with a pressure suit. Once suited up, the figure enters the air lock to the shuttle. Then all vision is lost to the smoke. A hatch activates, opens and closes, then a great rush of gasses. The smoke is now evacuated through the fist-size hole in the ship's hull as the last skin ruptures. The stricken crew momentarily freeze, then are torn into a vortex of blood and tissue by the ravenous vacuum.

A launch countdown has started and the computer-constructed voice, offering cancel options, falls on dead ears. The smaller parts of the of bloody tissue extrudes through the puncture in the craft's wall, and vaporises into the void.

Outside, in the darkness of space, the SBS falls away from the great silver ship. The gas-jet engines fire and accelerate the craft past EarthlabOne, and on towards Earth.

'Redman, can you hear me?' screams Major into the space station microphone, 'Get out of manual... Red. You must be able to hear me... I can hear you breathing. For Christ's sake, answer.'

He calms slightly, and then continues, 'SBS Orion – Redman, I assume you can't talk... you've still got your helmet on. For God's sake plug yourself into the unit. Don't think about it, just do it – basic training.'

A voice interrupts over the Lab's radio. 'Help me, for God's sake – I can't fly this blasted thing.'

Major is shocked, 'Hal? ... Harry? ... Is that you? Oh, my, God... what's happened?'

'The bloody thing blew up,' Harry's distraught voice answers, 'that's what happened. A bomb, get it? Another blasted bomb. Just get me out of this, you bastard.'

'Harry, just listen, there's no time for argument, the craft's in manual. If you're in the command chair you'll have just one screen, right? If it's blank, that's good. If not, don't worry–'

'It's blank, it's blank.' weeps Harry, hysterically.

'Okay, all you have to do is take your pressure-suit glove off and place your hand flat on the screen, so it can read your palm print. Then speak these words into the open mike: 'LIFT MANUAL OVERRIDE.' That's all you have to do, then we'll have you. That and plug in your umbilical... Hal... Harry, CONFIRM.'

'Okay okay, you bastard. What have you got me into? Okay, I'm trying.'

Inside the SBS, Harry is sitting before one of a pair of control stations. He is wearing a heavily armoured pressure suit with its bulky self-contained life support system backpack. He is desperately trying to negotiate the umbilical, which, after a few attempts he plugs in just below a single screen. A red glow now appears through the porthole, accompanied by the bang and clatter of the heat shield shedding redundant ceramic skins.

'I'm plugged in, but... damn, it... damn, it.' screams Harry in panic as he fiddles with his glove's wrist-joint. 'I can't get the sodding, buggering glove off, Christ. The blasted thing won't–'

'Slowly, Harry. Take your time.'

'Got it, I've got it. Right, taking my glove off. Be ready, I'm... WHAAAAAAAAT!... What in Christ. Oh my God. You crazy bastard... what have you done to me?'

As the glove comes away, there before Harry's eyes is a great grey carbonised plastic hand. He grabs wildly at his helmet, twisting and pulling, wrenching it out of its seal. He holds it before him, viewing himself in the gold foil visor. The reflection that looks back at him is all the more hideous being distorted in the convex glass of the dome. Grey, synthetic Harry hurls the helmet away and rips at the heavy suit, tearing the umbilical from its socket. He rises and stumbles across the cabin, ripping and tearing at the equipment, in blind panic. Major's voice screams from the radio to no effect.

On EarthlabOne Major turns from the monitor screen to Rena, 'Dear God, we've lost him.'

'Okay what's the predict,' says Rena, fighting to stay in control, 'What will happen to it without our guidance?'

'Without instruction, the craft's computer is trimmed to put the SBS in the best possible Earth trajectory.'

'Where will it land?'

'Almost anywhere... It'll try for US territory... that's its best.'

CHAPTER SEVEN

The stricken shuttle enters Earth atmosphere and starts to slow and cool. Computer directed drag-chutes pull it into a low glide over a frozen landscape. It hits snow and ploughs through to solid rock, tearing the craft in two. One half explodes into a ball of fire, the other continues on and comes to rest, out of range of the exploding fuel tanks.'

On the crippled Junairo3, the cancel option has now passed. A final countdown is spoken to the deserted command-deck. Already the SAMs have repaired the hole in the cabin wall and they scurry away, their work completed. The hull fills with mist as the atmosphere is once again restored. Harry's body lays in his sealed blood-smeared coffer, held in a cryogenic, dreamless sleep. On EarthlabOne, Major, Rena, and Hamish, plus various crew stand on the command deck. A grim Commander McQueen studies their faces as they look at her expectantly, hoping for news.

'We can't regain control of Junairo3.' She gives the information without emotion. 'It's about to blast off and if we don't tug her clear she'll knock us out of orbit. I'm sending an SBS to take her out.'

'There may be survivors, for Christ's sake,' growls Major.

'The order stands... it's my decision. You work on regaining control, but I must save the Lab.' Major nods reluctant agreement.

Rena slams the microphone angrily into Major's hands.

'What about Harry for Christ's sake? Concern yourself with the living, damn you. Do something.'

'Okay okay, you handle the shuttle, I'll concentrate on Junairo3… you have the authority. Well? … Jump to it.'

'The shuttle is down.' shouts McQueen, 'North America… coast… Alaska. We're still tracking for the beacon.'

Rena gives a sigh of relief, 'Right. I'll take an SBS, Earthside. Is one ready?'

McQueen nods, confirmative, 'Yes, it's fuelled.'

'Okay,' says Major, 'But you'll take Hamish.'

'No fucking way?' growls Rena, giving a defiant look, 'You said I'm to handle this.'

'You take him… that's final. He'll co-ordinate the search agency base, you handle it the field. Now, will you get to it.'

She gives a damming look to Hamish. He looks back with total indifference.

A whirlwind blows out of the north. The grey man – Harry in proxy – looks up at the sky. He moves in startled jerking actions. In his shocked delirious mind he is unable to comprehend the whirring blades of the search helicopter, the explosion and shock having dimmed the human faculty. Now only animal logic reacts. Grey Harry instinctively starts to flee, his newly-won wolf pack following close behind. His new body, being much stronger and fleeter than a normal man, leaps with animal ease over the icy terrain and on toward the frozen sea.

Onboard the helicopter, Rena perches in the open loading bay viewing the icy terrain through night binoculars. She shouts to a crewmember through her helmet radio-link. 'Don't lose him, he's making for the sea. Head him off. —Why is he running, he's clear of the wolves? Why?'

The chopper has now manoeuvred in front of grey Harry and is trying to flatten him with the downdraft. But his manufactured body proves too strong, he won't be beaten back. He runs onto the ice-covered sea, leaping from float to float. The water between is a churning torrent of small, chunks of ice, which can no longer support his weight. He sinks and ice-floats close over his head.

'Oh God, he's gone.' screams Rena, 'Get down there... Circle. Oh, Harry, I'm sorry.'

As the search helicopters makes a final pass, the wolf pack, now gathering at the edge of the ice, howl in unison. Below the surface, through the crystal clear water, Harry hears the dull deadened crunching of the colliding ice-floats. Over this he hears the muffled howling of the pack. His grey hand stretches out before him in the icy clear water. Animal instinct compels him to swims toward the howls of the pack, and to their moving shadows on the surface. He sees the wolves silhouettes cast by the moonlight. On the surface, the baying animals direct their howls to the movement just below them. An eruption of ice and white-water forces upward as grey Harry rises. The wolves momentarily scatter then regroup to greet their leader as he staggers onto firm ice. He takes one last angry look at the sea then runs off into the forest toward a distant curl of smoke coming from a small log-built hut. The pack follows.

After peering through the one small window Harry makes for the door of the little shack. He pushes it off its hinges with animal ease. The pack scatter and run off, bewildered at their leader's invasion into man's world.

Inside the shack, Harry finds a plain room. A log fire is just smouldering, but still with enough heat to hearten his exhausted, confused body and mind. After a few moments warming himself, he rebuilds the fire from a box of logs at the far end of the single room. His movements now becoming more co-ordinated, he manages to open a cupboard, wrenching the door off its hinges and grabbing at the food inside. After eating his fill he places the rest of the food on the floor in front of the fire and curls up around it like a dog and sleeps a troubled sleep. In his nightmare he is a gigantic wolf. He is bounding slowly through a surreal sky, a vortex of fast moving cloud changing intermittently into a desolate snowscape. Now the snow and cloud merge into an incomplete panorama, no one image claiming dominion.

The wolf, a creature of a dual faculty neither man nor beast, bounds through the snow in loping gait. All around are the dark remnant images of beings, the flotsam of humanity. His dream is an amalgam; animal and the remnant of his human psyche. He is in the Badlands.

In the distance, he sees the ethereal figure of a man. It is the body-proper of himself, petrified into stillness, set rigid by invisible bonds, his hands extended as if about to take hold of the bounding wolf. There are other figures lurking in the void, some vague and shadowy, some flesh and blood. There are figures moving in and out of the mist. All the time they are listening, watching with unseeing eyes. Harry – seeing through wolf's eyes – closes on the inert figure of his own body. As he is about to touch, the figure dissolves, and in its place is the grotesque form of Redman. His face is a morphogenesis of horror, his eyes moving under a covering, a continuous skin. The lids are not defined nor separated. He sees only through a single Cyclops eye set in the centre of his forehead. His teeth and tongue also move hideously under the same cowl skin. In his hand is an old-style fire axe with blade on one side and spike on the other – an exact replica of the one Harry had killed Kate Ottman with. Redman seems about to strike. A distant voice, slowed to an almost inaudible drone, calls out. Redman momentarily looks to the voice direction, then turns back to Harry and raises the axe. The voice calls again. Redman backs off into the void. The voice now slows to normal speech – a man's voice.

'Come out and show yourself. I'm armed, so make haste.'

The dream degrades back to the hunter's shack. The words make no sense to grey Harry, only the threatening tone. A thin beam of light falls into the room, followed by the barrel of a rifle.

'Stand still, there,' the voice calls again, 'I can see you. If you're friend you have nothing to fear.'

As the dark shape of a man enters the room grey Harry leaps – the rifle fires simultaneously. Harry lands in a heap. The huntsman, throwing himself backward, had loosed off a single round, smashing through Harry's side and ricocheted off his titanium pelvis, tearing out a massive gout of flesh on its exit. – Blackness.

Grey Harry opens his manufactured eyes. The blurred shape he sees sharpens into the huntsman's face. He tries to move but cannot. He is bound with rope and lying on a bed. The man standing over him speaks:

'I don't know what you are, Buddy, but if you don't keep still you're gonna bust out them stitches I just sewed into ya – I've tied off two arteries where there should have only been one.'

Harry snarls. The man continues, 'No, don't thank me.' Harry snarls again and pulls at the bonds, the hunter smiles, 'You don't understand a single word I'm saying, do ya? Well, they're coming for ya, whoever/whatever you are. You hear me?'

Harry growls, and tries to bite the man.

'Ooowee! You're one mean cuss. Anyways, you'll sleep now. I just gave you the mother of a jab... you should be out cold now. You ain't getting a word I'm saying, are ya?' says the man, shaking his head, 'Goddam strangest critter I ever did see.'

As Harry stares, the drug starts to take hold. Delirium again, the man's face suddenly dissolves into the Cyclops face of Redman. The axe comes crashing down. In his dream, grey Harry rolls away and runs off. Redman follows, thrashing wildly. The great wolf bounds alongside Harry, now in his proper flesh-and-blood body. The two figures dissolve into each other, an incomplete thing in an incomplete dreamscape.

The hunter's shack appears in front of Harry. He hesitates for a moment, then enters. Outside he can hear Redman's muffled shrieking voice coming towards the hut. A glance around the room reveals no other exit. Harry turns back to the door. A figure steps into the light. It is not Redman but the silhouette of a small boy. A momentary flash of light and Harry sees the boy's face. Another flash of light and the dream evaporates to sharp real daylight.

His blurred vision clears and he recognises Rena's face. Harry is lying in a hospital bed, tubes and wires appending every orifice. Rena is sitting beside him, holding his great grey hand. But now his synthetic body looks different. No longer does he have the heavy goggles set in the gargoyle head. His eyelids and face closer resembles the features of Harry proper. He opens his cosmetic eyes wider and looks about him. He tries to rise but falls back. He lifts his hand to his vision, and then looks to Rena, considers, then speaks.

'Rena, I saw him… I saw my son.'

She grabs him, gathers him up in her arms and cries out, 'Harry. Oh, Harry. We got you back.'

Harry looks about around his bed. Major, Hamish and a medic are standing in the ward. On hearing Harry's words, the medic rushes out.

Major walks to the bed and looks down at him. 'You've been through hell, son, but you did great – We want a full account–'

'Damn it, Major,' yells Rena, 'that can wait.'

'Of course… when you've rested, Harry, but first Rena has got to explain a few things. We'll leave you two alone.'

Major nods to Hamish and walks out of the ward. Hamish shrugs and follows. Rena waits until they've gone, then takes Harry's hand again.

'Harry, do you understand your predicament?'

'Yes, I think so. Will it stop me seeing my son?'

'No.'

'How long have I been here?'

'Three days lost, and a whole week in hospital, in educed coma.'

'I saw him, Rena, my son, in my dreams. I really did. '

'Things are moving, Hal. You can see your son as soon as you feel you can stand. This afternoon, if you like.'

'Yes, I like – but not like this,' he holds up his grey, synthetic hand.

'Yes, well... keep you hands in your pockets, or wear gloves. But we've...' she stops, not quite knowing how to explain, 'Let's just say we've prettied your face up a bit for the occasion – we couldn't do everything, not enough time. We've removed the eye protector and fitted optic lenses, your own colour. And we've cleaned up and done a few cosmetic chores... revised a few things. Your wound is mended... sorry, healed. That's the good thing about neurosynphysis... it's easy to manipulate and repair.'

'Why? What about my real body, it must be intact, the golden bowl? It... I must be... alive.'

'Yes.'

'Where is it, still on Junairo?'

Rena looks away as she answers 'Yes. We have contact. Junairo is about to around Mars–'

'WHAAT,' yells Harry, leaping up in bed. 'Oh my god.' You've got to stop it.'

'We can't Hal, it's set on course.'

'It's got no bloody shield.'

'There's no crew on board... they're all dead... and it's locked in manual. It's impossible to stop, Hal. We need the planet's pull to turn it... elementary physics. But there is a shield, and it's twice as powerful as the first, and it's directed around the coffers. We're certain the entity can't harm your body.'

'Certain? Is that the best you can offer? What's wrong with convinced?'

'You must see, Harry, your uncle's theory is sound. It enters through the human mind... your mind is here. Believe me there's no danger. When it rounds the planet we're convinced we can bring it in.'

'Convinced? What happened to *certain*? Joke... sorry.

What a mess. —I had just climbed when all hell broke loose. How I came to locate with this...' he holds up his grey hand, 'I don't for the life of me remember. It was total confusion. All I was concerned about was getting a suit on. —What about Redman?'

'We're not sure. His body is still on Junairo, but his systems don't feed back. We can't tell if there's life or if his shield is functioning – we know yours is, or you wouldn't be here.' She looks worried again. 'I've got to tell you Hal... I think Redman is dead. That is we hope he's dead, because if he's not, he's lost out there, in the Badlands. We're certain he was climbing the cord when the bomb ripped into the ship.'

'BOMB? What the hell's happening, Rena? Who? Is it the Russians? If it is me they're after, they've bloody well got me now, haven't they?'

'Maybe the Russians, or personal. That or there's one of them still here.'

'We should have tested every–'

'Done...' interrupts Rena, '... this morning. I had a numb butt for an hour.'

'Everybody – with my syringes?'

'No, not with your syringes – Everybody that was there: Major, Hamish, everybody. – Your syringes are with you on Junairo.'

'How about the Agency, the others?'

'What part of 'everybody' don't you fucking understand, Hal? We've used a bucket of novocaine... sorry. It was done in groups, everybody that was there was done, nobody was missed, nobody. Even the goddam cleaning staff. All negative.'

They fall to silence. 'Personal?' says Harry at length. 'You think someone is after me?'

'You said it first, Hal. You've made some dangerous enemies.'

'You mean the Mafia set the bomb? Never... not for this... Not for a bloody gambling dept.'

'Well... did you know Major set the Feds onto them? Perhaps it's revenge for that, they've done worse for less.'

'Bloody hell, the bastard. How about Hamish, perhaps he's in their pay?'

'The Mafia, I don't think so. I don't think it's Hamish... Could be, though, revenge for Rex. Anyway, I'm asking Major to assign him to other duties. He admits he's pissed with you, Harry. And he'd loved for that huntsman's bullet to have blowed your synthetic brains out. But he says it's not him.'

'I'm not so sure – I mean, he would say that, wouldn't he?'

'It's not him, Hal, I'm certain... convinced. Leastways he's out of it now. There must be someone, or something else out there.'

'Have there been any more bodies since I've been... away.'

'We don't think so... maybe. If it's feeding, it's covering its tracks.'

Harry attempts to get up. He stands but is still giddy, 'I'm out of here... I'm a sitting duck... a fish in a bloody barrel. How is Barney?'

'We did him too. Negative.'

Harry tries to leap up, 'I didn't mean that. Are you mad? He's just a baby. You can't shoot a baby full of blasted novocaine.' He staggers and is forced to get back into the bed.

'Everybody Hal, everybody that was there. Barney has had more tests than all of us put together. All negative, but we can never be sure. He could be a mutation. He could have got to Alfie or Ami. Think, Harry.'

'You are mad. What about you? If it was Barney, he could have got to you. It could be you. You could have brought that bomb into my office, and you have access to Junairo3... It could be you. Revenge for your father, because my shield didn't work.'

They stare at each other, grey Harry is unsure of her.

'That's it, Harry,' says Rena, not yielding, 'now you're thinking. We can't leave a stone unturned. You had a tangle with a British bookmaker. Do you think – '

'No, I don't think so... and that was a long time ago.'

'Still... they have long memories.'

'I don't think so.'

'Have you upset any other organisations: like the Freemasons, the Teamsters, the Mormons, Illuminati, L. Ron Hubbard, the Magic fucking Circle? – Christ, Harry, you're a goddam liability – Joking, sorry.'

'Please try to stay with it. And don't swear... no, I don't think so.'

Rena gives an, 'I'm not so sure' look. 'Okay, did you have any rival for Rose... any jealous lovers?'

'No,' says he, indignant, 'I was her first.'

'What do you mean, her first, her first what?'

'Her first... Her first and only lover. She was, you know...'

'You know, what? – You don't mean...?'

'Yes.'

Rena looks at him in amazement. 'No kidding, you were the first. Jesus... really?'

'Yes, really.'

'Okay, did she have any rival for you?'

'What is this? No?'

'Did she have any enemies?'

'I don't think so. There was some bad feeling with Kate Ottman… but I killed her too – you know about that.'

She gives a pathetic look.

'Don't bloodywell look at me like that. It could be Major… you said 'no stone', right?'

She gives a more pathetic look. Grey Harry takes her hand in his great paw and holds it to his cheek. She gives a worried smile.

'There's something else, Hal. I might as well get it all out in the open. We want you to consider something.'

'We?'

'Major... wants you to consider relocating with your body after Junairo3 rounds Mars.'

'WHAAAAAAAAT! You must be bloody joking. Tell me you're joking. You're not joking are you?'

'Just consider it.'

'Okay, I've considered it… BLOODY NO.'

'Hopefully it won't be necessary, but we have got to crack the computer, only you or Redman can give access.'

'No.'

'All you'd have to do is touch the screen, and we're in. Don't worry, with luck, we'll probably crack it before then.'

'Probably –With luck – is this the best you can offer?'

Rena ignores him and continues, changing the subject. 'So, if you're feeling up to it, we can see your son. It's only a short distance.'

'Now you're talking.'

'But first we're going to lighten your skin with theater make-up. What do you think?'

'Great. Will you do it?'

'No, we've got an FX man from the movies, he's the

one that oversaw your facial surgery.'

'Am I grey all over?'

'Have a look, big-boy,' says Rena with a cheeky smile.

Grey Harry looks amazed and annoyed at Rena's good-humour. 'Oh, good... I'm glad you're not too upset.'

'Com'on Harry, don't get shitty. It's not for ever. Have a look.'

Rena winks. Grey Harry is dumbfounded at her amusement. She nods at him, 'go on, look.'

He gets the message, lifts the bedclothes and inspects. 'Bloody Nora. Wow, have you seen this?'

'Seen it?' says Rena, still smiling, 'I commissioned it... part of the amendments. Now, be a good boy, and you may get to try it out.'

'Does it work?'

Rena leans over and touches him – he reacts. 'So far... so good,' says she. Grey Harry gives her a prudish look.

'Do you mind?'

'Come on, Moby Dick. It's hard for me, too.' She gives a chuckle.

Grey Harry peeks under the bedclothes again. 'You can say that again.'

CHAPTER EIGHT

From grey Harry's look of 'maybe this isn't so bad', to pink Harry half obscured under the brush and latex of Clive. Harry tries to speak:

'I–'

'DON'T TALK,' screams the manic FX man. 'Sorry, Lovey... please, pleeeease. don't talk, I'm doing your mouth.'

Harry grabs the unashamedly gay man's hand, stopping his fussing and fiddling. 'Listen to me, Matey, you've got a photograph of me there, and you haven't so much as looked at it once in half a bloody hour. I want my son to see his father, Harry Mandrake, not some ponced-up blasted, Frankenstein's monster. Now, look at it.' He thrusts the photograph.

Clive brushes the picture limply away with his powder-blusher. '*You* look, Lovey.' He lifts a hand-held mirror into Harry's vision. A pink Harry stares back, for all the world the real Harry – bigger and heavier, but Harry Mandrake to a tee. Harry's grey hand comes into the mirror's view, about to touch his cheek.

'DON'T TOUCH IT,' screams Clive, 'Sorry. Pleeeease, don't touch it. It isn't dry yet.'

'What's your name?'

'Clive, Clive Newman.'

Harry opens his dressing gown. The pink skin stops at his neck, the rest still dull grey. 'Is it waterproof, Clive?' says Harry, 'Because I think I'm going to cry, it's so bloody marvellous.'

Clive swishes about, accepting his praise overgraciously. 'Yes, it's waterproof,ish – but, well, you're not going to kiss anybody are you? ... Are you?'

'Just do the face and hands, the rest will do as it is, ha ha.'

Two hours later pink Harry is sitting in an armchair, soberly dressed and with a look of nervous expectation. He sits, hugely ridiculous, wedged between the chair's normal-sized arms. He considers a moment then stands, removes the two bottom cushions and pushes them behind the chair, out of sight, then sits back into the well of the chair, directly onto the rough undercover. He doesn't look so huge now. He wriggles and pushes himself further into the upholstery's void. He looks almost normal size.

After a few moments, a little boy toddles in, finger in mouth. He stops when he sees pink Harry. Son, and father in proxy, stare across the room. Harry's composite eyes flood with animal tears. Rena follows into the room and stands behind the little boy.

'God, he's the living image of Rose,' says Harry, 'You're right, Rena, he's nothing at all like me. Just like in my dream, just like Rose.' He leans forward and speaks to Barney. 'Hey, little man, how about a hug for your big Daddy?' Little Barney is stricken with fear and wonder.

Rena holds the boy's hand and moves him towards Harry. 'Come on Barney, say 'hello Daddy', like we've been practicing.'

Barney blinks and remains rigid.

'You come to him, Hal.'

Pink Harry stands and takes a step. The little boy's finger drops away from his mouth to allow him to cackle into laughter. Harry, in mid-step, is a huge tortoise... that's what Barney sees. Harry standing and looking ridiculous, the armchair resting half way up his back, lifted off the floor and wedged around his massive hips. He wriggles and wobbles, all to Barney's delight. The chair drops away. Harry picks up the laughing toddler.

Rena clasps her hands together, 'He likes you, Harry. He likes you.'

'What's not to like?' says Harry, with a grin, 'don't you know, all little boys love doggies.'

As pink Harry stands holding his son, two other people enter: a middle-aged dapper man, and an attractive, blowsy mid-fifties woman.

The man gasps, astounded, 'Bleedin' hell, Harry, what 'ave they bin feeding you on?'

Pink Harry is moved again to weeping. He speaks, choking back the tears. 'Alfie! Christ in heaven, where have you been? You're supposed to look after me... Look at me. Where the devil have you been?' In his outburst, Harry frees one hand from his son, walks the three paces to his old friends and grabs them to him to more floods of tears.

'They wouldn't let us near you Harry,' says Alfred, apologetically, 'I tried... I was going to the newspapers, then they offered a deal: If we backed off, me an' Ami got Barney. And you would be classified 'under observation', not 'committed' ... that was part of the deal. And...' He hesitates.

'And... go on. 'And?'

'Well, see, they wasn't sure if Rose.... sorry, not Rose... if the thing had got to you. You know, they wasn't sure if you was one of them.'

'And...?' repeats Harry, knowing there is something

else.

'*And*... they wanted you banged up till after the second Mars expedition, then they'd let you go. That's what they told us, Harry, honest. Did I do right? And they said if we get you to help them now, you know, with the space bloody rocket-ship an' all, we can take Barney back to Edinburgh, and you'll join us when it's over. Did I do right, Harry?'

Ami butts in, 'It's not his fault, 'Arry, we did what we thought best for Barney. We still don't understand about Rose.'

Harry pulls Alfie to him and kisses him. Alfie pulls away. 'Leave it out, silly sod—CHRIST ALMIGHTY. your face.'

As Alfie pulls away, a piece of latex make-up strips off from Harry's lip, revealing the grey texture beneath. Harry quickly covers that side of his face from Barney with his hand. He gives a moment consideration, turns to Rena. 'Okay, if you don't crack the computer I'll climb the cord back to Junairo3. I have to get my body back, I can't bloodywell have this... bits falling off all over the place.'

'Good. We've got two days free, Hal, what do you want to do?'

'A bit of a touch-up,' says Harry, with a smile. 'You know, put on a face, as you ladies say, then the zoo I fancy, what else? What say, Barney?'

From pink Harry's smiling face to a snarling inhuman mask of terror, accompanied by a savage roar: The lion's den. The beast, woken from its slumber by the scent of Harry, leaps to its feet and bounds across its cage and up to the bars to where the little group of spectators, Harry, Rena, Barney, Alfie, and Ami, stand gaping. Harry and son are at the Zoological gardens, Central Park. The family group moves on, to meander their way through the many walks of cages. People staring at the giant Harry, made all the more bizarre by little Barney perched on his shoulder wandering arm-in-arm with Rena... she, by comparison, now dwarfed. Alfie and Ami follow cautiously a short distance behind.

Small animals flee in terror, larger ones bare their teeth ready for fight or flight, all to the delight of little Barney. Just a normal, happy cybernetic, hybrid family.

Rena and grey Harry – his make-up almost gone – sit side-by-side facing Major. Hamish, by his master's side, eyeballs Harry. The old hatred is still there.

'No apologies, Harry,' says Major, 'You must see we had no choice.'

'Thanks for nothing.'

'We couldn't just keep you in quarantine indefinitely, we had to have a reason, mainly for our own people's morale. You must see... we didn't know if it had got to you. Now we know you're human we–'

'Human, HUMAN? For Christ's sake look at me, do I look bloody human?' No one speaks for a few moments. Hamish, with a sardonic half smile, breaks the silence.

'We didn't crack the computer, Hal. You're gonna have to climb the cord again to relocate with your body. Only you and Redman's palm prints are recognised by Janeiro's computer, so it's your ride. Redman's lost, so that just leaves little ol' you.'

'In that case, I'll just *find* Redman... he can bloodywell go.'

'Don't even think about that, Hal,' says Rena, alarmed at the thought of Harry going back to the Badlands,

'We've lost two more people trying–'

'Goddam it,' interrupts Major, angrily, 'that's classified information.'

'Damn you and your classified fucking information.' Spits Rena. 'Harry has a right to know.'

'That bastard has no rights to anything,' growls Hamish moving towards Harry.

Harry takes a step forward to meet him. Hamish's hand slips inside his jacket. They glare at each other. A stand off. Major moves between them.

Harry looks away to Major, 'I think I saw the two people out there... in the Badlands, and I think I can get them back. Who were they, who else did you send?'

Rena now moves toward Major, grabs him and screams into his face. 'Don't let him do it, you fucking maniac.'

Hamish pulls out a pistol and levels it in readiness. 'Back off.'

Harry moves to protect her. She lets Major go and turns to face Hamish and the levelled gun, 'You really think that is necessary, you fucking moron?'

'Yeah,' says Hamish, smiling, 'with a little luck your fruit boyfriend might think about helping you, then I can waste him.'

She steps between Harry and Hamish. 'Don't even think about going, Hal. First, they sent John MacKay, and when he got lost, that crazy bastard let Helen Cassidy try to find him.'

'Well, that's who I saw,' says Harry.

'Don't do it, Hal. In your state of mind you'll get lost too. That bastard doesn't give a damn for any of us.'

'Helen... I saw her, Rena.'

Rena ignores him. Calming slightly she turns to Major. 'Let's concern ourselves with the business in hand. If

Harry doesn't go back to Junairo3 we're going to lose her, we can't regain control.' She looks to Harry, 'God knows how long your life support will last. The state of the damage is incalculable without computer... could be a year, could be a day.

'A bloody day,' says Harry horrified.

Rena takes a step towards Major. Hamish raises the gun again. She leans her hands on Major's desk, keeping her distance. 'This meeting is over.' She points to Harry, 'And if he is going back I want him to get some rest. We'll continue this on EarthlabOne,' She indicates to Hamish, 'And I don't think we'll need him... you said he was out.'

Major turns to Hamish, he speaks loud so all can hear. 'Do you mind missing this, Son?'

'Hell no, Major. I got *important* things to do.'

'Okay. And for God's sake, put the piece away.'

Hamish gives Harry another look of bad intention as he puts the gun away.

Rena now lies naked in Harry's arms, her pink living flesh contrasting with grey synthetic tissue as they make love. They are travelling in the passenger cabin of SBS Orion. Unnoticed through the porthole the planet Earth slips away and EarthlabOne now fills the aperture and, in the changing sunlight moonlight earthlight, the entwined couple tumble slowly with the added pleasure of weightlessness.

Contrariwise, in the brightly coloured command room, Major, Hamish, and commander McQueen sit in angry discussion. After a few moments the commander slams her hand on the table and stands.

'This is not finished Major, not by a long way – I demand an official inquiry.' She turns and leaves. Major is quite for a moment, Hamish is about to speak. Major raises his hand.

'No. No more talk. You go, too. I don't want another scene with you or that Limey fucking... werewolf. I've had it with him... he's going to split the whole goddam sack.'

'What have I been saying?' says Hamish.

'Right. Just go. But stay close, and don't let them see you. I don't trust that gutless wonder-boy now he's got muscle. Shoot him if you have to.'

Hamish nods in pleasured agreement, then leaves. As the door closes, Major takes a semi-automatic pistol from his jacket. He removes the magazine and checks the clip of blue-nose shells, then slips it back and slides a bullet into the breech. He puts the gun away and moves behind the desk.

The door hisses open to commander McQueen, Harry, and Rena. McQueen ushers them in, giving Major a damning look as she turns and leaves. Without waiting for them to sit, Major begins to speak.

'Okay. When you get onboard and locate, Harry, all you have to do is open your coffer manually. You know how to do that, I take it?'

'Don't antagonise him, Major,' says Rena, 'he's in a good mood, I don't want him to lose it.'

'I don't do antagonise – just fucking listen. Hal, after you've opened the coffer move to the command console, place your hand on the screen and say 'lift manual override' and that's it... nothing else. Get back in the coffer, seal it, then walk the cord and get back here, pronto. Is, that, clear?'

'What about the entity, them, there's no shield?'

'The Junairo is out of the danger zone,' says Rena.

'Are you certain?'

'I'm positive.'

Harry gives her a grey, damning look, 'Well if it's all the same to you I'd like to–'

'Is that fucking clear?' yells Major, 'Just touch the goddam screen.'

'Yes. But I want to try to find Redman and Helen, first. We've still got time. I've seen them out there.'

'No way.'

'Absolutely no way,' adds Rena.

'Look,' says Harry, pleading, 'all they need is someone to guide them. You shouldn't have sent them without my instruction. They don't know what to look for, they–'

'This isn't a goddam game we're playing,' yells Major. 'The whole Agency is riding on this. We are talking big money... half the fucking national debt. Forget the others.'

'You can't mean that,' says Rena, incredulously.

'Can't I?' He stares long into Rena's face, then looks away. 'Look, we had to send them... hedge our bets. If they fucked up we had to have a back-up.'

'Back-up?' says Harry, 'There's no one else left.'

'They knew the risks when they volunteered, everyone and everything is expendable, myself included... Everything, except the Agency. If you won't go, I'll send someone else. There's a goddam queue of fledglings wanting in on this project. One of them can locate with the other synthetic and try to bring Junairo3 in manually. So yes, fuck the others, and fuck you.'

Grey Harry hurls his words and himself at Major. 'You crazy bastard, they're out there in a living hell, and you'd risk more?' He grabs Major by the throat, lifts him off the floor and shakes him. Rena screams at Harry to let Major go.

A shot!

Harry's synthetic face contorts as a bullet smashes through his body. He flinches and stops. Standing behind him is Hamish, his great silver pistol still smoking is levelled at Harry. There is a huge bloody bullet exit-wound in the middle of Harry's back.

Hamish fires again. Harry recoils, releases Major and staggers.

Another shot.

It is Hamish that now spins and falls to the floor, his pistol clattering across the deck, his gun-arm shattered. He attempts to pick it up with his left hand, Rena stamps on the hand as he grabs at it – his fingers crumple. She replaces her own pistol under her tunic and picks up Hamish's magnum, and points it back at Major.

'You bastard, you said he was out.'

Grey Harry falls to his knees. Rena runs to him and holds him in her arms. 'Harry, oh Harry. You've got to walk the cord.'

'Help me, Rena... help me, I'm dying.'

'No. Your synthetic body is dying, Harry,' says she, tears flooding her eyes. 'You've got to leap... you've got to get back to Junairo and relocate. Don't think about the distance, that's what you said, just follow the tracks, follow the silver light like before when you were a boy. Do it now while you have the strength... Harry.'

Harry lies still. She lays him down – the two pulses flicker then stop, the synthetic's life extinguished.

A white wolf bounds slowly through the netherworld landscape of the Badlands, leaving bloody paw-marks in the snow. The inert body of Harry proper appears in the distance. It stands completely motionless. The wolf bounds up to it and leaps at the spectre and they fuse together into one, the amalgam body of wolf and Harry. After a few moments he animates. He takes stock of his surroundings and starts to move off.

Suddenly, the grotesque remnant of Redman steps from the mist, swinging his axe. Harry parries the blow and grabs the weapon by the handle and manages to wrench it away. Redman screams muffled abuse under the taut, cowl skin. Harry ducks and clubs Redman to the single eye in his forehead with the butt of the axe. The eye bursts and dark bloody fluid gushes forth. Redman backs away, screaming under the membrane. His image degrades away into the mist.

Harry recovers and walks on towards the bright light ahead of him, a ray of light stretching out to infinity. It is the silver cord. A woman is lurking in the mist. Harry beckons to her but she shrinks away back into the swirling void. He turns and walks on into the mist, following the light to Junairo3 just visible at the end of the silver cord blowing out its massive vapour veil.

CHAPTER NINE

The flight deck on Junairo3 is dim misty and silent. Harry's ethereal body manifests through the decking. Translucent, he slowly strides the dozen or so steps to the bank of coffers. He reaches up to the top tier. In one coffer his body lies in suspension, its paired coffer is empty. On the lower level is the suspended body of Redman, this coffer is intact but covered with debris. In its pair is the inert body of a synthetic.

Harry's spectre, now merges with his flesh-and-blood body inside the coffer. Immediately the scene brightens and leaps into life as dream gives way to sharp real-time. Bellowing klaxons demand attention. A pair of SAMs, still tinker, still assessing and reinforcing the earlier repair. Inside the crystal coffer, Harry's body starts to move. He touches a control and the seal breaks. The coffer momentarily fills with mist as air mixes with gas, then the mist is gone. Harry proper rises and makes his way to the simple control console: a row of switches and two display screens, both blank. He sits at the console and places his hand on the screen and speaks.

'Lift manual control.'

Immediately Rena's face appears. She shrieks out joyously, 'Harry. Oh, Harry, I didn't think you'd made it.'

'Okay, now what?'

'Just put your lips to the screen.'

'My lips. Why?'

'Because I want to kiss you.'

He does, she does – a zillion miles apart, they kiss. After which Rena opens her eyes. She looks past Harry. Her joy is quickly replaced by terror.

'Listen carefully Harry,' says she pulling away from the screen, 'I want you to get back in the coffer... then I want you to purge the cabin of atmosphere.'

'Yes, but–'

'Don't talk, act. Just get back in and seal it up and walk the cord... your body will sleep till we bring it in. But before you pull the plug I want you to reset the shield: divert it just to your coffer... Do you understand, Harry? Confirm, Harry?' she screams, 'DO YOU FUCKING UNDERSTAND?'

'Yes, yes. Bloody hell, what's wrong?'

'Nothing. Just do it.'

'You said the Junairo was out of their reach, Rena. Tell me what's wrong, what's changed? I'm not moving until you tell me.'

'There's no time, Harry, please God, just do it.'

'I mean it, I don't move...'

'Okay. Look about you... what do you see?'

'Nothing, just a smashed coffer and debris.'

'Any blood?'

'Strangely enough, no.'

'Well, there fucking ought to be... blood and tissue from the four crew, all over the goddam place.'

'Okay, okay. So what's happened to it? The SAMs must have cleaned up.'

'If the SAMs had cleaned up they'd have cleaned the debris too. I think it's been *used*. Dear God, It's been used. There was only one human body, Hal, Redman's... it needed more human flesh... It used the scraps. I only hope there wasn't enough. Now move.'

Harry scans the cabin. The top coffer is open, waiting. The two coffers below have the synthetic in one and Redman's body in the other. Both appear sealed and dormant, debris scattered all around, no blood or tissue. Harry makes his way slowly up to his coffer.

Rena's voice rings around the cabin. 'For god's sake get in. Quickly – MOVE!'

Harry, shocked into action, makes a dash to the coffer. Suddenly, a shower of shattered crystal throws out into his path stopping him dead in his tracks frozen with fear. Redman's grotesque, distorted body emerges from the shattered coffer. He has the same mutated face, still carrying the wound from their Badlands encounter, the Cyclops eye smashed and blinded. Redman's own eyes move under the cowl skin. His teeth now gnaw at the inside of this fleshy membrane... biting, devouring his own tissue to make an exit for the bloody, slimy tongues. He places his hands up to his sightless mutant face, his fingers penetrating deep into the skin either side of his nose and ripping away two huge flaps from his face, exposing his bloody gargoyle eyes. Slowly he lifts himself from the coffer just by his arms, like a crab, his malformed unfinished legs dangling useless. He stares across the cabin at Harry through blood-filled eyes, tongues leaping from his mouth. Harry forces himself to move. All this time Rena is screaming orders at him from the screen, all falling on deaf ears.

Harry's hand finds the edge of his coffer and he tries to pull himself into it. The tongues lash out around his body and pull him back, stopping him. One of Harry's hands reaches into the coffer, but his feet lose their grip in the slime, and he falls to the deck. What he'd been fumbling for now falls gracefully in the quarter Earth gravity, just out of his reach. It bounces slowly onto the deck and settles. It is one of his syringes, the other, still in its holster, hanging over the edge or the coffer. Harry reaches backwards and fumbles for the fallen syringe, his fingers just touching its tip. The tongues pull and the syringe rolls out of reach. Redman's head tips back like a great swaying sea anemone, its Gorgon hair striking out at Harry's struggling body and pulling him ever nearer to the drooling orifice, grabbing at arm, leg, neck, body, holster – A stray tentacle unwittingly grabs the leather strap, mistaking it for a part of Harry's anatomy, and pulls at it. It comes free. Syringe and case fling across the short distance towards Redman. As they hover momentarily in low gravity, a perfectly aimed lunge from Harry's free foot sends the razor sharp needle clear through the leather case and into the soft, mutated flesh. The syringe empties into Redman's neck. He gives a violent convulsion then, as the burning novocaine ravishes, a gurgling rush of blooded matter erupts through his gash mouth and nostrils. Harry is free. He crawls away on his elbows and rolls over, props himself up on his knees and looks to the screen. Rena looks down at him from the console.

'It's not over, Hal.'

Harry's face, smeared with toxic blood-slime, has a look of bewilderment. 'Not over?'

'No, Harry, behind you.'

Harry rolls away and looks back over his shoulder. The second synthetic, the one from the top tier sharpens into focus. It is now standing back behind the coffers, watching.

Harry looks to the screen. 'Who is it, Rena?'

'I don't know. They're all gone.'

Harry looks to the cabin floor for his other syringe, it's

not there. The synthetic moves and lifts its great, grey hand. The syringe looks tiny in its huge fingers. The fluid voids onto the cabin floor and the syringe is discarded.

'Who is it, for God's sake? Redman? John? Is it you, John?' He stares at the huge grey synthetic face waiting for its answer.

'Lo' Harry.' It is a woman's voice, low, sexy and ludicrous, it coming from the great male synthetic.

'Helen! ... Helen, is that you? You made it back, thank God.'

'I followed you Harry, and I led Redman. I guided him to his body then I located with this,' she indicates to her synthetic body and looks at the screen, 'What do you think, Rena? Does my butt look big in this? D'you think Rose would approved? Have I got balls, or what?'

'Watch yourself, Harry,' screams Rena, 'Something's wrong. Make your way to the lab section and take a cutting lance and protect yourself. Something's wrong.'

Grey Helen moves and narrows the angle of retreat. Harry backs away toward the console.

'I'm going to kill you, Harry, as painfully as I can.'

Harry looks to Helen then back to Rena. She has a look of dismay mixed with fear.

'Move, Harry,' shouts Rose, 'MOVE!'

Grey Helen steps towards Harry. 'You still don't get it, do you, Harry? I volunteered to walk the cord to find John, just to avoid your little experiment with the novocaine. I knew they wouldn't risk waking my latent body with a jab.'

'You planted the bombs.'

'*Bloody bingo.* ... Yeah, I planted the bombs.'

'But, why? Why?'

'You killed Rose, and you killed Rosette. I loved them

both.'

Rena screams 'Get out Harry, NOW.'

'You took Rose away from me, Harry. If it hadn't been for you she would have loved *me*. Then Rosette came to me, after you hurt her arm, I looked after her and loved her. She looked after me, she made me one of them, and she loved me. And you killed them both.'

'But in this form, you shouldn't have the urge to kill'

'Oh, I have the urge to kill... I hated you before and after, Hal. For the record, Rose did love you, and still you killed her. You see, we can love... even gay gods can love.'

Harry, stricken with terror, turns to the screen and mouths two words to Rena. She reads his lips, 'Golden bowl.'

Helen looks to the screen, to Rena. The grey face makes a smile and speaks. 'The minute you leave that screen, Rena, your lover-boy is dead... but I want you to watch.'

Rena, sitting at the console on the EarthlabOne flight deck, turns away from the macabre drama. She gestures with her eyes to Hamish, lying injured, propped up against the decking where he had fallen just out of ship-camera vision. He has seen and heard everything.

Rena muffles the mike and speaks softly to him, 'Go to it, the bowl, you know what to do.'

'No, I ain't doing shit... I'm bleeding.'

'Go to it, God damn you.' She kicks his pistol back to him across the cabin floor. It hits his foot. He just looks at it.

On Junairo3 Harry is cornered. He manages a dash for the other side of the coffers, but the agile synthetic is on him. She grabs him by the arm and with the slightest effort dislocates it at the elbow. Harry screams.

On EarthlabOne a door bursts open under Hamish's shoulder charge, He is now moving down a corridor crashing, running and staggering from loss of blood, the big silver pistol clattering along the metal walls cradled in his smashed arm.

He charges at the airlock doors – they automatically open and he hurls himself on down the winding causeway in sub-gravity loping gait.

Harry is now on his knees. Grey Helen is standing over him, holding his dislocated lower arm. She snaps it like a twig. Harry screams again.

'That was for hurting her arm, Harry. How did it feel, anything like this?' She snaps his upper arm. He gives a muffled whimper.

Rena calls to him from the screen. 'Don't pass out, Harry. Hold on... she'll kill you if you pass out, pleeease.'

Grey Helen laughs, 'You hear that, Hal? Don't pass out, *pleeease.*' She starts on his other arm.

Harry calls to the screen: 'The bowl. Break it... break it. Aggghhh.' His other arm snaps. Helen looks to the screen.

'Say goodbye, Rena, Harry's got to leave us now.'

Rena looks away, unable to watch. Hamish reaches the sickbay. He tries to point his weapon in his broken hand, but the pistol clatters to the ground.

Helen's great grey fist lifts up above her head and is about to bash Harry's brains out. She hesitates – her human face momentarily superimposes the grey, manufactured face, as if to relish the final moment in person.

On EarthlabOne, Hamish struggles desperately trying to manipulate the pistol in his mangled hands. He is trying to level it at one of three coffers, at Helens' head through the crystal lid.

The grey hand starts its way down towards Harry's head, simultaneously the pistol fires. The bullet smashes through the glass and into the head of the inert body. A momentary flicker of the eyes as the grey hand finishes its blow, slightly off target. It glances off Harry's head and imbeds into the metal decking.

Hamish is shocked and horrified as he watches the body of Helen erupt into a gurgling mess of blood and slime. The novocaine from the blue-nose bullet ravages the mutated flesh.

Harry's eyes flicker. He sees Rena, out of focus, pleading from the console.

'Harry, Harry, answer. Harry, please listen. Don't die yet, please. Silvercord, you've got to climb again, Harry. Just one more time. Pleeease, Harry.'

Harry's crushed head rolls over and he looks, one-eyed into the screen. He manages to whisper. 'Bye Rena, I love...' The one eye shuts as the SAMs start to crawl to him with life support equipment.

A swirling cloud of white vapour and a flare of light: Harry stands, a gossamer thin translucent image. Before him is a brilliant flood of white light. He is drawn to it. Silhouettes of people stand inside the light with welcoming arms. He walks towards it. He hesitates, turns and moves away. His eyes now focused on a lesser light. The silver cord is now visible through the churning mist, stretching out to infinity and to EarthlabOne, there in the far distance.

Harry turns away from this light, too. He walks slowly off into the mist and fades away to nothing.

CHAPTER TEN

Rena joins Hamish in the sickbay. The big man is sitting against the cabin wall and a medic is treating him. A little way away are the three coffers, one of which is covered with a blood-smeared sheet. In the second is John's inert body. Rena stares into another coffer, the one containing a synthetic bearing the grey features of Harry – his forlorn hope of relocation.

She diligently studies the grey, frozen features. She speaks to it as if it heard her words. 'Come back Harry, think of Barney. We all love you... don't give up... please. Please, Harry.'

Major enters the room.

Rena calls again to the lifeless grey hulk. 'COME BACK, HARRY, DON'T LEAVE ME.'

'He's gone,' says Major. 'I'm sorry.'

'Damn you, you murdering bastard.'

'I am sorry, believe me. We've got the read-out from the SAMs, they've got him in suspension, but ... the bowl is broken, I'm afraid.'

Rena puts her head in her hands, silently weeping. She stands and raises her hands to her mouth screams, 'COME BACK HARRY, PLEASE COME BACK.'

As if in answer to her words, the crystal coffer shatters. Rena and Major are showered with shards, as a fountain of jewel-like fragments tumble lazily in reduced gravity. Grey Harry ascends like a Phoenix rising from the ashes, a Lazarus back from the dead. He crashes through the lid of the coffer with a great bellowing cry of triumph. 'Bloody bingo.'

Major Staggers back, covering his head with his arms. He yells, aghast, 'It can't be. The golden bowl – we broke the fucker.'

Rena gives Major an incredulous look, then grabs at the synthetic, embracing it in her arms, glass and all. 'Harry. Where the hell were you? Where did you go?'

Grey Harry looks back down the sickbay, to the other coffer. 'I went back for a pal.' He points a finger. Rena follows two his direction to the far coffer as it opens. John MacKay sits up and greets Harry with a glib hand flick, GI salute.

END BLOC TWO

> *O world. O life. O time.*
> *On whose last steps I climb,*
> *Trembling at that where I had stood before:*
> *When will return the glory of your prime?*
> *No more – Oh, never more.*
> Shelly

BLOC THREE – DEMIGOD

Author's note:
Two short points before I conclude my trilogy.
—Point one: the enigma of frogman, Commander Lionel (Buster) Crabbe: George Medal, OBE, RNVR. After the WW11, Crabbe was solicited by MI5 to look at the hull of the Russian cruiser Ordzhonikidze, which was bringing Kruschev and Bulganin to Britain. He was never officially seen alive again. Wild rumours circulated about his disappearance, one suggesting that the Russians had abducted him, another that Crabbe had defected. The Crabbe affair, which should have become open to the public scrutiny in the 1980s under the 25-year embargo rule, is to remain sealed until 2057.
—Point two: the debacle of the Liberty Bell 7 space shot and the infamous exploding hatch bolts. Although all mission elements were successful, the flight was marred by the loss of the Liberty Bell 7 capsule, following splashdown. Gus Grissom made his own way from the capsule and out into open water. The actual post-splashdown transcripts do not back up the popularized book and movie accounts depicting Grissom panicking and accidentally blowing the hatch. Grissom, in fact, requested extra time inside the stricken capsule to take down certain instrument readings before being recovered. Subsequently, and 'conveniently', as most conspiracy theorists would say, Grissom heroically died in an accident on the launch pad along with two other astronauts. Please read on –

Who called you forth from night and utter death;
From dark and icy caverns, shattered and the same forever?
Who gave you your invulnerable life,
Your strength, your speed, your fury, and your joy?

Coleridge

CHAPTER ONE

1952 – SPACE, NEARING EARTH ORBIT: Into a star-spangled heaven comes a gigantic eruption of gasses. Insignificant compared with the deep indigo void these gasses continues to expand. At the centre of the gaseous vortex a single speck of solid matter appears, growing ever bigger as it hurls onwards. It is a crude, metallic cylinder that seems oddly familiar, more like a World War II submersible than the perceived spacecraft. In its wake follows a deadly shroud of propulsion material as it careers on into an Earth re-entry trajectory.

The craft enters Earth's mesosphere and the massive power unit abruptly changes angle. Pusher retro-plates redirect the explosive thrust, causing the missile to decrease velocity. Now sparks and fire trail as it hits the stratosphere. Lumps of debris break off from the crude heat-barrier and streak away leaving fiery contrails, the steel hull of the craft glowing white-hot, just a few degrees from meltdown.

OUTER HEBRIDES: A rumble of what seems like thunder. A growing roar, and now a high-pitched whistle. There is a glow in the heavens, growing to a huge fireball and leaving contrails of smoke through the night sky as it crashes to Earth.

1956 – PORTSMOUTH HARBOUR: In cloudy

moonlight three frogmen busy over their equipment. Two of the men make the sign of the cross as they pull on their rubber garments. The third man, the leader, draws deeply on his cigarette and blows a coil of blue-white smoke into the cold night air, then throws the stub into the black water. He looks at his two companions. They are waiting for his lead. He quickly pulls on his upper garment then steps into a rubber cummerbund, rolls it over his lower body and unravels it at his midriff, securing the two-piece rubber diving suit. He then slips on his breathing apparatus and smiles:

'Right, you ugly buggers, let's get our feet wet.'

The second man nods, 'Buono fortuna... Dio dalla nostra parte.'

The leader shrugs, not understanding, and gives a caustic, 'Oh yeah? That's easy for you to say.'

In the pale moonlight, they slip into the water and make for a large cruiser. The great ship's name is just visible in the gloom, *Ordkhonikidze*. A signal-light flashes from the ship. The lead frogman returns the signal with his torch. The other two frogmen stop short of the ship. The leader turns and points to his watch, raises five fingers and then disappears into the blackness.

Moonlight shimmers off an expanse of green/black water, illuminating the occupants of a small fishing vessel conducting a burial service. The mourners are assembled on deck around a body ceremoniously covered with a Union Jack. At length the deceased is tipped into the sea. It slips slowly from under the flag, revealing the legs clad in a tight-fitting rubber suit then the top half of the body, headless and with little substance. It gracefully disappears into the black water, barely creating a ripple. An hour later, on dry land, these same people are now gathered at a graveside. The gravestone simply reads, 'COMMANDER CRABBE.' Nothing else is on the marble slab – just empty black.

A vivid monochrome yellow invades empty blackness: An age has passed. This formless diffusion of light gradually sharpens to a field of golden daffodils swaying in surreal mazy motion, tossed by a light breeze. There is an air of timelessness in this pleasant, country vista. An entwined couple embrace amid beautiful flowers. An immeasurable period elapses before they break away from their lovemaking. It is Rena Lansavitch and a bronzed Harry Mandrake.
    Harry lays his head in Rena's lap and looks up at her and sighs, then closes his eyes again, dreamily.
    'Bored?' says Rena. Harry doesn't answer. After a long pause, she continues. 'What do you consider the most important of all historical mysteries, Harry?'
    After an equally long pause, Harry lazily opens his eyes and condescends an answer. 'Oh... let me see... I've got it... the true location of Camelot. I've always held the belief, as did T E Lawrence, that it is the ancient Roman town of Camulodunum, the modern-day Colchester. *Camulodunum*, like *Londinium* – London, get it?'
    She gives him a contemptuous look. 'You're not too smart, are you? I like that in a man.'
    'Hey, steady on, old girl... it's only a theory.'
    'Movie, Hal... What's it from, and what year?'
    'Oh... easy, just seen it – 'Body Heat', circa eighty-four.'
    She smiles and they fall to silence once more. After a

little while, she speaks again. 'Are you bored?'

Harry considers, but still doesn't answer. He closes his eyes again and lets his thoughts wander, the strong sunlight flaring through his closed eyelids. After an incalculable lapse of time this light degrades away back to black.

1983 – DEEP SPACE. Into blackness a monitor screen suddenly flickers to life. The weak light cast by this screen dimly illuminates the flight-deck of a huge spacecraft. It is the interior of the rogue starship, Junairo3. Bloody tissue and glass scattered from the shattered coffer, litter the once orderly flight deck. It is the grizzly remains of Redman, blown apart by a syringe-full of novocaine.

The great centrifugal spindle comes into view. It is motionless. Of its four appended coffers two are shattered, one holds the inert body of a synthetic, and the other the comatose body of Harry.

A silver cord of light appears. From which Harry's levitate body manifests and enters through the ship's gunnels. The two onboard lanterns flicker to life and Rena's face peers from the monitor. Her voice echoes around the cabin.

'Hal, I can't see you, but I assume you're there and can hear me. I've activated the control deck, if you're not back in twenty seconds I'll join you as planned… so leave the cord cast for me.'

Harry turns to the screen and waits. The twenty seconds tick past and Rena's transparent body steps from the silver beam of light and onto the flight deck. Harry greets her and they immediately begin surveillance of the craft. Rena studies the illuminated dials, committing as much to memory as she can. She stands engrossed, going over and over the same information, fixing it into her mindset.

Harry stares down at his battered body in the coffer, still blooded despite the SAMs life-recuperation systems. It is the flesh-and-blood body of himself sealed in the crystal coffer, held in suspension in an oxygen rich electro-plasmatic atmosphere, breathing once every two minutes – the breath of the near dead.

Rena joins him and they both stand looking down at the cadaverous remnant. Rena speaks in a thin echo voice.

'It's not good news, Hal.'

'What, the read out of Junairo3, or of me?'

'Well... both. The Junairo is volatile but she's good for a month, then she'll slip orbit.'

'Which way?'

'Deep space, hopefully... probably. Then... God alone knows. As for you, your broken arms are healed and you'll stay like that indefinitely.'

'Indefinitely?'

Rena nods to the body, 'A hundred, two hundred years if nobody touches it... the blood's artificial and recyclable and the systems are self-sufficient, so is the ship.'

'A hundred, two hundred years?'

'You're doing it again, Hal, repeating what I say. I've told you about that.'

'Sorry.'

'And stop saying sorry, for Christ's sake, Harry.'

'Sor – Okay okay, bloody hell. So...?'

'So, there's no ballistic rocket fuel left, just the atomic-reaction engine, and we dare not use that – It'll pollute the place before we've ever set foot on it,' she sighs, 'Come on, we're finished here.'

'Okay. Put out the lights.'

'Why?'

'To save bloody power, I may need it. After two hundred years I may want an extra day. Okay?'

Rena laughs and they both walk off, following the silver cord of light and leaving the cabin in darkness.

CHAPTER TWO

A dimly lit cavern opens. In the half-light, someone's eyes travel along its tubular ethereal structure. The walls are repetitive in formation, neither solid, man-made, nor organic. A sudden sharp turn and the cavern wall opens like a virtual living thing, allowing access through a swirl of confused images. These images now cohere into solid objects, holding form long enough for study. It appears to be a scene from the distant past. John MacKay's voice gives narrative over a kaleidoscope of fluctuating images:

'It is clearing, Hal... Jeeezus. is this all in my mind? It's like watching a movie. A group of white-robed holy men are addressing a gathered crowd standing in a great temple colonnade. This is fockin' cosmic, WoooWee.'

'Calm down John,' Harry's voice interrupts, 'just tell us what you see.'

'Okay okay. There's a bunch of suckers wearing sheets, togas or something. Like, this is crazy. Now there's a crowd gathering. Wooo,weeee! There's some fancy chicks... sorry... a group of young women... and some men, surroundng a man who appears to be the leader.'

Rena's voice now interrupts. 'Could you describe the surroundings, John.'

'What? Oh yeah... Hey, what the... Two huge lions on a wall in blue coloured ceramic tiles. They stand guard either side of the great bronze doors. Jeeezus.'

'Tell it as you see it,' says Harry urgently, hardly able to contain his enthusiasm, 'Just tell it.'

'My God, Harry, I think I know this. This is ancient Mesopotamia. Shoot, I know this. I've seen this in'a museum. Jes' Harry, I know this. Hey. The leader guy is coming forward. He's a great big son-of-a-bitch warrior prince by the look of his clothes and weapons. He's about to address the crowd. This is crazy, I know what's going on without anyone telling me.'

'Okay okay, so what now?' says Rena desperate for information.

'I just know, Rena. How the hell can I know this? ... I just know.'

'Okay, you just know,' says Harry, 'Just keep telling what you see.'

'Right Hal. Okay, just gi'me a moment. There's been a sign, an eruption or something on the Red Star... that's Mars. The big guy is holding up his great studded club, he's pointing to the distant mountains, and...' he hesitates.

'And, and...?' prompts Rena, impatiently.

'Hey, these mothers ain't stupid. They know what's happening. A shit-load of volcanoes erupted on Mars two months ago – half the goddam planet. It dumped a whole mess of material into space, most of it formed a ring around the planet.'

'A ring?'

'Yeah, Hal... These suckers have called it the 'Enkidu Girdle'... 'Enkidu' – Sound familiar to anyone?'

'Not to me.'

'Does to me,' says Rena. 'I think... Never mind, carry on, John.'

'That's where we/ they are now, their last bastion of survival. Something on the planet surface, some mineral released by the volcanoes was killing us/ them/ us/ them. What the hell am I saying? Them, them.'

'Don't worry about it now. Just continue.'

'That's easy for you, Harry, but–'

'Believe me, John, we will sort this, I promise.'

'I don't think so, Hal. This thing is for keeps, it ain't just for Christmas. It can't be sorted.'

'Trust me, John,' adds Rena, 'I will sort it, one way or another.'

'Anyways... some of this matter broke free of the planet's pull and now the Earth is being bombarded with meteors. Most of which burn up, but a great mother crashed into a mountain a couple of hundred clicks away and this big son-of-a-bitch warrior, this Hercules-type dude, is going t' go check it out – Yep, he's gone.'

John is quite for a few moments. Harry prompts again, 'What's happening now, John?'

'Hey, he's back. He's returned – that was quick. He looks older and drawn, looks like he's been through some great trauma... I know how he feels.' John throws up his hands and sings out *Looney tune* style, 'Da-da-da, da-da-da. *That's all, folks.* – I don't know how I know that, it just is.'

The metaphysic cavern closes and John is left sitting in a small ward in the Carnegie airstrip hospital. The room is bare except for some chairs and a small table. John is bound to his chair with metal restraints.

He blinks nervously as he speaks. 'What are you going to do, Hal? —You said you'd sort it.'

Grey Harry, his face now finely remodelled into his old likeness, is standing looking at John. Behind his back, he is holding one of his syringes. John leans to one side and sees it. He is terrified.

'Go ahead. Empty the goddam thing, I don't want this shit. Now I know how Cameron felt.'

Harry squirts a small jet of the cloudy yellow fluid into the direction of John's arm. A few drops splash onto his skin, John recoils in agony. The skin blisters.

'Christ. What are you doing? Get it over, Harry. For the love of God, finish it.'

Rena douses a jug of water over the arm, washing the burning novocaine away. Rena and grey Harry, stand looking down on the heavily tethered John.

'Sorry, John, I had to be certain.'

'Certain. I'm a goddam alien, for Christ sake. Believe it.'

Rena moves forward. She puts her hand on John's shoulder and speaks tenderly to him. 'I'm so sorry John… but we have to know what–'

'What's it like? I hear voices, Rena… I got memories I never had before.'

'Memories,' says Harry, 'How far back, John?'

'The distant past, primeval… Hell, I don't want this Hal. —I'm an American, I don't wanna be no scum-sucking alien. Get it over with.'

'Sorry, I can't do that.'

'I think you'd better, Hal. I also have the urge to kill you… both… it's all I can do to fight it off.'

Grey Harry pulls up a chair and sits next to John. 'Listen to me, old man, we do understand… it won't come to this.' He puts the syringe down onto the table.

John stares at it. 'Do it while you still can, for God's sake.'

'No. We need you. You say you've the feeling to kill me… how come, I'm not human, I'm synthetic. What exactly do you feel? Do you feel you have the strength to kill me?'

John looks Harry up and down and considers his huge

synthesised frame and muscular arms. He looks into Harry's eyes:

'Now that you know – you know that I'm an alien an' all – you've become a threat. It's the instinct to survive. Believe me, Hal, I could rip your plastic carbon fibre throat out and the only way you could stop me is jab that needle into me. Is that what I've got to do to make you do it?'

Hold on, John, there is a way out. We use a synthetic. You walk the cord, just your human psyche. Then we seal your body, entity and all, in suspension then you locate with the synthetic.'

'No, way. I ain't no zombie, neither.'

'It's working for me, John. Am I a zombie?'

John, attempting a smile, 'You're a Brit, all Brits are zombies, didn't you know that, Harry?'

'I take it that was a joke?'

John's smile fades, 'Leastways you got a chance, Hal. When they bring in J3, you get your body back. You got hope.'

'It is working, John,' says Rena, 'Harry's not a zombie, I can vouch for that, he's more of a man now than he ever was.'

'Hey. Steady on,' says Harry, deeply offended. 'I did okay before. I don't recall any complaints.'

Rena rolls her eyes, as to say, 'Christ, Harry, just for John's sake.'

John shakes his head in disbelief, 'So you're saying I spend the rest of my days as fockin' Plastic Man? – I don' think so.'

With a mighty effort, John rips one arm upwards, breaking the metal restraint straps. Harry leaps back out of his seat. Rena levels a pistol at John's head. As it cocks, six blue-nosed novocaine-filled bullets revolve in the chamber. John just sits forward and puts his head in his hand. He speaks, as he looks sideways to the syringe on the table.

'Christ. Both of you get the hell away. Just leave me alone, leave that syringe or give me the gun or a blue-nose, I'll bite on it... I'll do it myself.'

'Hold on,' says Harry trying to appeal to John's patriotic conscience. 'You say you're an American, right? Well, you can still be of great service to your country... I'm sure you wouldn't want to deny them that.' John considers. Harry has almost cracked it. 'What I suggest now is, John, that we all walk the cord. I'll explain out there in the Badlands, it'll seem different out there... I have a plan.'

John gives a look of despair, and shakes his head again. This time, in grudging agreement.

The mist-covered landscape of the Badlands extends to infinity. Within it are silhouettes of moving shapes, fragments of lost beings. Through this churning vapour, a beam of silver light constructs. Rena and John manifest, and stand side-by-side looking down into a deep abyss. Clouds drift by, some in accelerated speed and some in slowed motion – time seemingly ungoverned in this netherworld.

A great white wolf bounds in timeless gait, moving out of itself and leaving second degrading images trailing behind. The second image is Harry, gossamer transparent. The Harry image now dominates that of the wolf – his proper flesh-and-blood body. He walks to join Rena and John and they stand together in the swirling mist.

'So, John, how do you feel, now?'

'I don' know, the same I guess. You look great. One thing I don't understand, Harry, how come with you changing size an' all, how come you clothes still fit?' Harry gives him an, 'are you serious' look. 'Jesus, Hal, I hate this place. I don't want to kill no one no more, but I still got the memories. Don't ask me how come – if the entity is gone, how come I still got the memories?'

'It seems the memories stay even after the catalyst entity is gone... memories are memories, I guess.'

'Anyways, I don't have the feeling to kill. At least that's gone. Hey, we got t' move, there's danger here. The bogey man is here.'

'I don't understand,' says Rena, 'If John has lost the urge to kill, why didn't Redman?'

'Give me a minute,' says Harry, 'I'm sure I can work it out... maybe because he was under Major's murdering influence.' – As Harry speaks his image degrades away and the great white wolf is superimposed, the two images have equal presence, the wolf is now dominant leaving just a thin image of Harry.

'Yeah, I think he made a deal with Major,' says Rena, giving him a wearisome look. 'I think they meant to kill you, Harry. Something Major said when you made it back with John... when he saw you alive,' she mimics Major's voice, '*The golden bowl, I thought we broke the fucker.* Anyway, that's what he said.'

As Harry speaks he resumes his former self, the wolf image degrading away, 'I don't believe it, Rena.' The wolf image again starts to return as he continues, 'I think–'

'WILL YOU STOP FUCKING DOING THAT GODDAM WOLF THING.' yells Rena turning on him with exasperated fury. 'Jeesus, Christ. Harry, you really piss me off sometimes.'

'Sor,ry... pardon me... bloody hell.' The wolf image immediately disappears.

'Believe it, Hal,' says John, 'Major can be one evil son-of-a-bitch.'

'Well, I'll be damned.' Harry shrugs, now believing.

'Damned? You are damned, man,' says John, smiling. 'That is a fact. You're in fockin' Hell. Whooo,weee!' he lets out a yell to help along his untimely stab at humour.

'Yes,' says Harry, non-plussed, 'that is funny, John... just. I laugh later, when I've got a little less on my bloody plate?'

'Are you serious, Hal? ... I never know with you.'

'When you two have finished pulling each other's pigtails, can we get the fuck on?'

Harry gives her a disappointed frown. 'You really must stop swearing, Rena, it doesn't help.' She gives a withering look. Harry continues. 'Okay, we move on, but first, we have got to lose this mist... that's where the bits are. I'm not sure if they're dangerous, but they sure scare the hell out of me.' He looks around at the lifeless terrain of snow and ash: leaden sky to swirling horizon, just bleak desolation, and the mist.

'God,' says Rena, 'it can't all be like this.'

John shrugs, 'I've spent a lot of time here, Rena, an' I ain't seen a bird, blade of grass nor a leaf in the whole goddam place, and I do mean God damned. How 'bout you, Hal?'

'Well, John, let me see. When I first walked, when I was a boy, I'm sure I saw–'

'Can we get on?' yells Rena, angrily interrupting, 'Christ's sake, we got priorities. Let's move away from these freaks, they make my skin crawl. Then I want some answers.' She walks off. John and Harry obediently follow, away from the incoming mist. They stop at the edge of the overhanging abyss. Rena begins to interrogate John again. 'Okay, what we want is your memories. But first I need to know, are you still on our side?'

'Fockin' A, Rena... I don't owe them scum-suckers a thing.'

'Okay, as long as we know. Now, they were here before, pre-history, Rose told Harry. So, if they were got rid of before, then what we want to know is how? Do you think you can pick up where you left off, John?'

John looks at Rena and nods. He sinks to the grey, lifeless ground, and squats. Harry and Rena do likewise. As John closes his eyes the cavern opens up to him. In his mind's eye he travels the tunnel once more, into the kaleidoscope of changing images. A picture coheres as John narrates:

'Bingo Rena, right on the button. So, the big guy's back – Hey, I'm getting good at this – Seems he's been away a whole year. Jesus, does he look pissed. He's killed his best friend or brother, whatever... something. Anyways, he's changed. Now they hail him as a King.'

'King?' says Harry.

'You bet ya, Hal. This sucker's hit the jackpot: 'Seer of all things' they call him... something like that.'

'King of what?'

'King, god... king of everything, something, I don't know... this mother is immortal. Hey! I know who he is. Shit, I've read this... this guy comes down to us as Gilgamesh. I've read this.'

'The Epic of Gilgamesh,' says Rena, 'Is that what you are saying, John?'

'Yeah, that's what I'm saying. Well, I'll be... That's it. Gilgamesh and Enkidu, I know I knew that name. They think a god has come to visit. Suckers... are they going to find out? Hey, it's going.'

John shakes his head to clear his thoughts. He is back to the Badland dreamscape and the tunnel closes. He shrugs and continues to Harry. 'I know it, Hal, all of it. They came in the distant past, when we were cavemen... they made modern man. Then for some reason, they died out. Now they've come again, like some goddam super virus, on the meteor from Mars, from the Enkidu Girdle. That's where they are now.'

'They're still there? How many?' says Rena.

'Yeah, it's still there, and they don't do numbers. We, Earth always supposed that liquid water was the precursor for life – not so, there ain't so much as a goddam drop on Mars. They/we just are, that's all, we just are... and we're a collective, an '*it*', until we do... well, you what we do. Anyways, Gilgamesh was the first... from the second time, that is, and he ate his brother... that's how they put it, and we know what they meant by that, yeah? And from him came the rest. Christ, Harry, they were the gods... Ye Gods. Harry. Get it?'

'We gathered that.'

'No, you don't get it, Rena. We/they were the gods, the Greek gods of old times. They say in every myth there's an element of truth. A race of Gods. Woooo!' He stops talking and closes his eyes again. After a few moments, he continues. 'Genesis six, chapters one to four: ''And it came to pass, when men began to multiply on the face of the Earth, and daughters were born to them, that the sons of God saw the daughters of men, and that they were fair: and they took them as wives, all which they chose.'''

'So they could breed with humans?'

'Yeah, but they begat humans, the only way we/ they

could make... themselves was to... Well, we know what they do.'

'But I need to know–'

'Sorry, Hal, I got to finish this first, it's all coming into my head at once and I can't stop it – *'And the Lord said, "My Spirit shall not always strive with man, for that he also is flesh: yet his days shall be a hundred-and-twenty years. When the sons of God came into the daughters of men, and they bore children to them." '*

'What became of the children, John?'

'Don't stop him, Harry,' says Rena. 'Go on, go on, John.'

John opens his eyes. 'That's it... end of sermon. It's as close as don't make no difference.'

'Could you expand on the bit about the children, John? I need to know.'

'Sorry, Hal, you'll have to wait till it comes to me. I can't control it yet. I do understand its priority.'

'But I desperately need to know.'

'Hey, this makes me a god, right? On your knees, jerks, show some respect, wooo,weee!' Harry smiles. But John's humour is short lived, he continues, dejected. 'Fuck it, Hal, I don't want to be no god.' He turns to Rena, 'I didn't know she'd done it to me, you know, Helen and stuff. I think it happened when I tied one on. See, me an' Helen had this big argument about her wanting to climb an' all.'

'Climb?'

'Yeah, Hal, you know, climb the board, surf. She was determined to walk the cord. I know you tried to stop her... she told me... and I'm grateful. Anyways, I went out on a bender, tied one on. When I got back I must have passed out.' He closes his eyes and immediately the event plays out to him: he is now inside Helen's apartment, the prelude to his drunken night. They are arguing. Finally, she hurls her drink at him as he disappears out the door. He is now staggering from bar to bar, now in a flashy nightclub. He stands drinking off a huge glass of bourbon, a woman sitting beside him. They leave together. Later, in her apartment he attempts to wham-bam her up against the wall. Amused, the professional hooker delights in John's naïve schoolboy lovemaking and resolves to enlighten him.

Still with his eyes closed, John views as voyeur to his own seduction, comic almost, featuring every position imaginable. His education completed they sleep, she naked, serenely beautiful illuminated in silver moonlight, and now the ravishing whore illuminated in scarlet strobe neon-light from an adjacent hotel sign, and now serenely beautiful again, in changing neon blue. John, lying beside her in a drunken stupor doesn't notice a fleeting shadow.

The neon light extinguishes as the first rays of morning sunlight illuminate the room. The hooker is now awash in blooded fluids. Helen is standing over her, fingertips delving deep inside her rib-cage, one hand either side pushing as if the fingers would meet somewhere between her heart and lungs. As the fingers retract, the woman's body corrupts into a mass of viscous slime.

Now Helen tenderly holds John's head, she weeps as she moves her hands slowly down to his neck, the fingers penetrating deep into his flesh. His body arches into a delirium of pain and ecstasy. The two corrupting bodies fuse together and, from the widening pool of offal, a new John emerges, resurrected. Helen, now also naked, clasps John's slime-covered body and they embrace. Her mouth covers his, pushing aside the blooded jelly covering his face. She and John lock together in frenzy and take their alien pleasure.

John opens his eyes and the image is gone. 'She must have done it then, when I was drunk, because we didn't argue no more. That was two days before Junairo3 went AWOL.'

'You mean you didn't know?'

'No, not right off, Hal.'

'So, when? How long have you known?' Says Rena, hardly believing.

'I don't know. Not long. I don't think she would have done it if I hadn't tried so hard to stop her surfing, do you Rena? D'you think Helen really loved me?'

'I don't think she did, John.'

'Shit. What a mess. I tell you, Hal, I want out.'

'How were they got rid of before?'

'How the hell do I know? It didn't happen in the same week. You're asking me to scan through thousands of years of chronologic history.'

'So what? Scan,' yells Rena, angrily. 'Or have you got something better to do? You just did it when you told about Helen, for Christ's sake.'

John shrugs, 'Yeah, that's right. I tell you I've had this shit.' He considers, turns to Harry. 'You realise you're in bad danger. You guys don't never listen. The bogeyman is here – yous two better get going, I'm staying. I ain't going back… not to be no goddam Mr. Spock.'

Rena puts her hand on his shoulder. 'You've got to come back, John. Christ, you can't stay here. If you break the cord you'll become part of the bits. You'll drift around, lost forever. You've got to come back and locate with a synthetic.'

'I ain't doing that, neither. I ain't going to be no fockin' Klingon.'

Rena hugs him in his despair. She whispers intimately, out of Harry's earshot, 'Come back, John. I promise I'll set you free when the time comes.'

'That time may have come already... look.'

The mist has moved in. It has all but surrounded them – the bits now move towards them as a huge body. The silver cord is momentarily lost in the swirling fog.

'Don't let them close the gap,' shouts Harry desperately.

'I can't see the cord,' yells Rena. 'What do we do, Harry?'

'God, I don't know.'

'The only way is over the edge... they've out-manoeuvred us.'

'Hope you can both swim,' says John as he prepares to jump, 'I think, hope, there's water down there. I'm fockin' over. —Wooo,weeeee!'

Before Harry can argue, John is gone.

Rena grabs Harry's hand. 'Okay, Hal, jump on three... One.'

Harry looks down over the edge, he shudders and looks away, 'What if there *is* water down there? I can't swim.'

'Two. – *Are you crazy, the fall will probably kill you?*'

They move to the edge as the mist is on them, things are pulling and grabbing at them. She looks at Harry, raises her eyebrows. Harry looks at her, puzzled.

'What?'

'Well?' She looks at him again, her eyes demanding an answer.

'Well, what?'

'What's it from?'

Harry rolls his eyes as he falls in. 'Christ sake, Rena. Butch Cassidy and the Sundance Kid, circa nineteen-seventy-four – Wooooo,hoooooooooo!' He grabs her by the hand and disappears over the edge, pulling her after him.

Rena screaming as she drops, 'THREEEEE.'

They fall slowly, like Alice down her rabbit hole, but not into Wonderland. They land in a Limbo wasteland, still in the land of the unbaptised forgotten and unwanted things. It is now a landscape of snow, slag, dirt and slate-grey permafrost. The only illumonation is from the disappearing mist, the light from the silver cord, reflecting off the patches of dirty snow. Harry and Rena survey the lifeless void. John is nowhere to be seen.

'Okay, there's the cord. You go back, I'm going to find John.'

'No way, you come back now with me, please. Let's get the hell out of here.'

'We need him, Rena, he's our only clue.'

'And I need you,' she snaps back. 'He's got some decisions to make. He may have decided to stay.'

'But I have to know what happens to the children… to Barney, you must see that, Rena. You go back, but I must find him. I'm sorry.'

Rena walks into the band of light, resolute there is no other way.

CHAPTER THREE

From a swirling mist the white wolf appears, bounding aimlessly in through dimness of the underground gallery. The spectre of Harry superimposed over it, matching its slow timeless gait. He stops and rests as he considers his surroundings. He calls to John, his voice echoing around the cathedral-like cavern, reminiscent of a scene from the opera Aida. Harry's mind wanders back to the time with Rena at the opera, the marvellous music, and the beautiful dancers. He is dancing, slowly gyrating in time with the music; he is dancing with Rena.

The music has now changed. It is much faster. Now he has a different partner, her face obscured by flowing hair. As her head tips back he sees that it is his deceased lover and mother of his son, Rose Hawkins. She smiles at him and they kiss, a long passionate kiss. As he pulls away, her hair falls and again obscures her face. The music is beautiful, exquisite. She tips her head back. It is now Rosette, and there is blood on her neck – a little trickle just below her ear. It runs down around her neck and into her necklace.

Harry gives a growling moan, desperately trying to break the hallucination. He screams out in terror, 'Naaaa naa. No, Christ, no.'

The apparition fades as Rena steps out of the blackness and takes his hand. 'You okay, Harry.'

'Rena. Dear god, I thought–'

'What you think, Harry, is what you get – remember,

that.' She smiles sardonically and, under his disbelieving gaze, begins to morph into the hell creature, Rosette, drooling tongues and blood-slime. Harry's eyes dart around for some way to flee. No way to the left, and no way to the right. In desperation, he looks upwards and contemplates the immeasurable distance to the surface.

'Tartarus.' John's voice booms out from the blackness, 'Same distance down as the Earth is from the Moon... so they say... so it is written. Did the Ancients know something we couldn't even conceive off, Hal, a different dimension?'

As John steps out of the mist, the Rosette image backs away into the shadows.

'Jeees, Harry, I'd pass my Classics degree now.'

'Thank God. She was just about to–'

'That's where we are, Harry, *Tartarus*. The chief God banished a couple of scum-suckers down here 'bout four K ago. The Circle-Eyed Ones... let's say the Cyclops, and The Hundred-Arm Ones... let's say the goddam Medusa. Any bells ringing, Hal?'

'I don't want to think, John.'

'Anything like this?' John is now dribbling saliva from a widening, grotesque mouth. The nightmare vision of Rosette, that was just a moment ago an apparition, is now incarnate in John. Harry is terrified, frozen to the spot in fear as John moves nearer to him. The tongues now leap out. Harry closes his eyes in dread. He can feel the hot breath on his face, getting closer and closer, almost on him, now almost touching. Harry is literally quaking with fear. The hot breath passes by as John walks on into the darkness, in pursuit of Rosette. Harry, gingerly opening one eye, stands amazed.

From the darkness comes the noise of ripping and

tearing, followed by an unearthly scream. Two figures roll out into the dim light, locked in deadly combat, awash with blood and slime, tongues slash and rip. They are lost into shadow again, continuing the fight in the total darkness. Another harrowing scream, then silence.

Harry, recovering from his fear-induced stupor, backs away from the blackness. A figure emerges. It is John, carrying the body of Rosette, which he dumps down at Harry's feet. John staggers and drops to his knees. He is injured, a great flap of flesh has been torn away from his side, exposing four blooded ribs. He rolls over onto his other side and fits the dangling flap back into place.

'Don't worry, Hal, this will be healed when we get back, that's the way it works – don't ask me how I know. It's only really real while you <u>think</u> it's real. When you start to hallucinate, it exists. Take a look.' He indicates to the bloody body at his feet. 'The bogey man/woman... it's whatever you think of... your worst goddam nightmares.' He looks at his side. 'Shit, Harry, look at the mess I'm in. When's this going to end?'

Harry studies the blooded body of Rosette. 'What are we going to do with it?'

'Don't ask me, Hal. You dreamed it up,' he indicates to his wound, 'This too.' He looks to the body. 'Bury it, or leave it as a warning to others.'

'Others. What others?'

'Well, that's up to you, Hal.'

'Are you saying that this is another of the rules of the Badlands?'

'Yeah. You wanna take my word for it or try again... What you dream, is what you get.'

Harry gets the message. 'That's what she said. —Can you walk, John?'

'You mean surf?'

'Well, no... Yes, surf?'

'Shoot. Why not, I can always come back. You keep your mind on the business, now. No daydreaming, right? You know what I'm saying here?'

'Yes, I know what you're saying there.' Through the blackness, the silver band of light stretches out.

John, holding the flap of flesh in place, smiles at Harry, 'What did you think I was going to do back then, Hal, give you a big wet kiss?'

'You're not funny, John.'

Harry helps John up, offering his shoulder as a prop. John gives him a strange, very suspicious look. His eyes follow down to Harry's midriff. A look of repulsion replaces the look of suspicion. John sniffs the air, pulls away and takes a couple of steps backward, a mixture of disgust and amusement on his face.

'Phew,wwheeee! You follow behind... I'll walk upwind if you don't mind. I hope *that* don't follow us back.'

Harry waddles on uncomfortably, his legs in parenthesis, trying to negotiate the unwanted package swaying in his underpants. 'You're not funny, John, not funny at all.'

CHAPTER FOUR

A flare of white light, turning gradually to pale blue. There are objects dancing in this light, tiny silver objects tumbling timelessly in slow motion and shimmering in a cloudless sky. A shower of tiny silver horseshoes showering a picturesque church. It is confetti descending a wedding group: The bride and groom being Harry's middle-aged housekeeper, Alfred and his lady friend, Ami. Among the wedding guests is synthetic Harry, in heavy make-up, and with Barney perched in one arm and Rena on the other.

Harry leans and gives Alfie a kiss on his cheek, 'Good luck, Alfie. Good luck, Ami. Look after little Barney – he couldn't have better parents. Good luck.' He kisses Ami, then Alfred, and then Ami again, then goes to kiss Alfred again.

'Give over, soppy sod,' says Alfred, squirming away, 'you'll smudge your lipstick.'

Ami grabs Harry and Barney in an embrace. She pulls back and studies Harry's huge frame. 'When you're settled, you know when you're right, he's still yours… he'll always be your son.'

Alfie whispers, aside to Harry, 'Bloody hell, if my old Missus could see us now she'd turn in her grave. Who'd have thought it, Harry? me in America an' getting bleeding hitched again. Who'd have – Umphhhh.'

Ami light-heartedly elbows him in the stomach: 'You'll turn in your bleed'n grave in a minute, using such language in front of little Barney... And I'm your Missus now...' she winks at Harry, then looks back to Alfred, '...Darling.' She grabs Alfred and gives him a big, lippy kiss. Alfred squirms, looks at Harry and rolls his eyes as to say, 'look what you've got me into.'

Ami lets him go and turns to Harry. 'We can adopt him now, Harry, legal like,' she nods towards Barney, 'They've said no one will oppose it.'

'And Major...?' says Harry, suspiciously.

'Sorry. Major is part of the deal,' adds Alfred. 'Don't worry, Son, I looked after your father and I looked after your uncle, when they was teenagers – God, those two was real buggers. Not like you, you was a miserable little git–'

'You watch your language,' interrupts Ami, 'What have I told you? And stop bullying Harry.' She smiles to Harry, 'He don't mean nothing, he's a bit nervous.'

At the wedding reception, Ami and Alfie lead the dancing. Harry and Rena are the first to follow. Coloured lights from the old-fashioned mirror-ball, dance and girdle the hall.

Outside, Harry waves goodbye to the newlyweds as they depart for a long-weekend honeymoon, leaving Rena and Harry with Barney for two whole days. Each hour stolen, to be treasured and stored for memory as possibly the only moments like these Harry will ever know: Rena bathing the little boy, Harry teaching him baseball then cricket, obviously the latter being the hero. Harry, batting, hits a huge ball, sending it out of sight and breaking the bat in the process, Barney crying, then both laughing. And so to bed: Rena and Harry reading a bedtime story, a page each in turn... Harry still reading on, long after Barney has fallen asleep.

Harry is now naked and looking ridiculous in his pink make-up – just head and hands, the rest of his body still grey synthetic, He is making love with, pink all-over, Rena. A very classy, very romantic, very Anglo-canine moments of sweet erotic foreplay.

CHAPTER FIVE

In the officious oak-panelled courtroom, Major slouches arrogantly. A semi-circle of other high-ranking military and civil dignitaries, sitting aloof and adjacent to fixed microphones, confront him. An unmistakable air of hostility hangs over the auditorium like a leaden weight. A court official calls out a name:

'Commander McQueen to give evidence.'

McQueen leans forward from her seat and speaks into a microphone in front of her. 'Okay, here.'

The adjudicator looks up from his many papers. 'Lizabeth Jane McQueen, are you commander of EarthlabOne?'

'Yes.'

'Remember, Commander McQueen, although these hearings are in camera, The Secrets Oath, nevertheless, still very much applies.'

'How could I forget?'

'Kindly give us your account of the space-tug lost in the ill-fated USS Junairo3, incident.'

McQueen gives a damning look to Major, and then turns back to the adjudicator. 'When Junairo3 went rogue I sent a tug to pull her clear of EarthlabOne. As they were taking it to deep space, the Junairo's engines activated and the tug let her loose. Then *another* engine fired... the tug was caught in the flash-flare... the second engine was nuclear. A reaction slow-burn atomic pulse-engine... I have no doubt about that. And–'

'You have no doubt of that?' interrupts Major, leaning forward into his microphone yelling at the top of his voice, 'What you mean is you have no proof of that? You have the tugs debris I take it? And you've checked for gamma radiation?'

'YOU KNOW DAMN WELL I DON'T.' McQueen spits her words back at Major, 'That craft was completely lost. The blast of that atomic engine blew it to smithereens and out of orbit, crew an' all… four of my people were lost, damn you.'

The official bangs his gavel. 'Let's have some order. Now, since you've interrupted, Major, what *was* the nature of this second power unit?'

'Christ, I can't tell you that. That's classified. —Who the fuck do you think you are?'

The adjudicator leaps to his feet and barks into his microphone. 'You'll moderate your language Sir and answer the question.'

'No, I will not, and no I will not. That's an official secret.'

'There is no national secrets enforcement here, this judiciary is in camera, you are a private company and your directors have demanded this hearing. The Russian shareholders have confirmed that their contribution to the Carnegie Space Fund was atomic engine technology. Now, answer the questions or you'll be in contempt.' He glares at Major, awaiting an answer.

Major stands and leans to his mike. *'Private company…* do you realise the United States Government is the biggest shareholder of that company? The President himself is involved, personally – 'Space Bonds'. The Junairo program was to put a man on Mars, to give America sole sovereignty… You know what the hell I'm saying here?

Sole sovereignty?' All Russia gets is long-leased landing-sites and facilities.'

The official bangs his gavel again and yells back at Major. 'Answer the question.'

Major smiles, 'I forgot the question, already – 'American sovereignty', that's my answer. No one had a hair up his...' he looks to McQueen, '... or *her* ass about that. And if it wasn't for that Limey saboteur, Mandyke, that's what we would have done.'

The official calms and continues, exasperated. 'Will you answer the question: What was the nature of this second power unit?'

'Okay, so Junairo3 had an atomic-reaction engine. Because we'd lost our window we had to, shall we say, 'fold' across space... for that, we needed nuclear power. So it laid a trail of radiation from here to Mars, so fucking what? What do you suppose the Sun's been doing for the past zillion years? The radiation is negligible.'

The hall erupts in angry voices.

'You will moderate your language Sir,' yells the adjudicator, over the din.

'The hell I will. —Do I continue?' The noise dies down. Silence. 'Okay. Now, it would take a team of Einsteins to isolate a single particle from that engine. It's only dirty if it's used in earth atmosphere or close orbit. Other than that it's like a gnat, pissing in the ocean.'

Uproar and the banging of the gavel, Major continues over. 'If you want to impeach for pollution, you need evidence.'

The hearing degenerates into chaos. As the chairman tries to regain order, Major collects his papers and starts to leave. An attendant marine bars his way. Major turns on him with hostility.

'Get the hell out of my way, Son,' he shouts, so all can hear, 'I out-rank you and everybody in this goddam room.'

The confused marine stands aside as Major walks on. Once outside the courtroom, Major marches off towards the exit. Rena and Harry, still in heavy make-up, bar his way.

Rena challenges him, 'God, you're class act. You made a deal with that pink-eyed dwarf, Redman. You incited him to kill Harry.'

'He didn't need no inciting. Same as I told that bunch of fuck-wits, you want to impeach, you need evidence. Now get out of my way.' He looks past Rena to where Hamish is standing. Rena's eyes follow his gaze. Hamish puts his hand inside his jacket and pulls it out empty. He makes an imaginary gun, points his finger and mimes a shot. He smiles and joins them.

'Trouble, Major?' says Hamish as he arrives.

'Trouble? Nooo, I wouldn't call it that, just a little annoyance.'

Rena looks at Hamish. She lets her eyes drop to his gun hand. 'How is the hand? You still this maniac's gunsel?'

'Gunsel. ... I like that. That's from something, a film, yeah?'

'Could be.'

Hamish remembers, 'Oh, *Maltese Falcon*, wasn't it? 'gunsel', I like that... Oh yeah, the hand... *here's looking at you, kid*.' He whips his hand in and out of his coat, this time he is holding his huge silver pistol. He spins it half a dozen times and gives a couple of side twirls then, in a flash, slips it back in his jacket, 'The hand's okay, I guess, thanks for asking.'

They stand staring at each other for a few seconds. Major beckons to Hamish. Rena and Harry move off, leaving Hamish and Major together.

'I want rid of that mongrel, once and for all,' say Major as he eyeballs Hamish. The big man is silent. Major continues. 'You hear me, boy?'

'I draw the line at murder, Major. I'm not an assassin.'

'It won't be murder.'

'I ain't no 'gunsel', neither.'

'It won't be murder. He's not a man, he's a goddam dog – shoot him like you would a mongrel. Call it a cull. Just get it done. Now, we leave for EarthlabOne in six hours. We've a date with Junairo3. Come through for me son, you won't regret it.'

Two hours later, Rena and Harry enter the airstrip hospital. The decor has changed since Harry was last here. The glass-panelled wards are gone, just stark partitioned rooms with attendants on every door. Some of these guards greet Harry and Rena some do not. There is a distinct air of hostility as they walk on through the corridors, passing guards and technicians and on until they come to the room where John is held in quarantine. The seated sentry nonchalantly touches his big peaked hat without the courtesy of eye contact. They ignore him and enter.

Inside, John is tethered to the chair with heavy metal bonds. He is reading a translation of Greek Mythology by *Robert Graves*. He speaks without looking up, 'It's amazing. So close, yet so far... from the truth, that is. Most of it is right, but it's naïve, as if the world facts have been recorded by the mind of an adolescent kid. I know this, me being a god an' all.' He finally looks up. 'Hey, my favourite people.' He holds out both his hands in greeting. As Harry is about to close the door, the guard from outside nudges it open and enters. As he does, he lifts his peaked hat from over his eyes – it is Hamish, with gun in hand. Once inside, he pushes the door closed with his shoulder; he speaks as he screws a huge silencer to the still aimed pistol barrel:

'Unfinished business, Hal... you got to answer for Rex.'

As Harry and Rena back away, Harry tries to reason. 'Wait... just hear me out.'

Hamish cocks the pistol, a row of blue-nose bullets tumble in the chamber.

Rena steps in front of Harry. 'For God's sake listen to him.'

Harry pulls her out of the way. 'Let me handle this.' He turns to face Hamish. 'Hold off for a minute, just hear me out. Did you *see* Rex's body?'

'Yeah, both bits.'

'Did you see his tongues?' Harry turns to John and winks, 'Did they look anything like this?' Harry nods, 'show him'. John obliges, widening his mouth. Hamish's eyes move to watch, the pistol does not. A single tongue leaps at lightning speed and coils around Hamish's hand.

–A shot.

A dozen more tentacles entwine Hamish's arm. The gun falls to the floor. Rena picks it up and points it back at Hamish. The tongues retract. John wipes the bloody slime away from his face and grimaces.

'You're gonna think this kinda crazy,' says he, slurping the goo from his lips, 'but I hate the taste of this stuff.'

Hamish recovers enough to speak. 'What the hell....'

'You said it,' says Harry, 'Hell, that's what Rex had become. We were friends that last day... friends. He promised nothing would happen to me as long as he was around. Believe me, it's the truth, Rosette had followed you, she got to Rex.'

Hamish looks from Harry to John, then back to Harry. 'How come you haven't killed him too, seeing as he's one of them?'

'Ah-hah.' says Harry, sarcastically, 'So... you haven't noticed the fact that he's tied to the bloody chair with steel bonds.'

Hamish rolls his eyes and shakes his head, 'Point taken.'

'Anyway, we have a deal,' continues Harry. 'We think we can help him. I'm sorry it wasn't an option with Rex... we're on a learning curve.'

Hamish studies Harry, 'You should know, Hal, you're bleeding... your shoulder.'

Harry puts his hand to his shoulder and inspects. He is indeed bleeding. 'Bloody hell. You shot me... again. Will you please stop bloody shooting me.'

Rena levels the gun at Hamish. 'You bastard, I should waste you here and now.'

Hamish shrugs, 'Do it.'

'Pity it wasn't me,' says John, 'Would have saved a deal of trouble.'

Rena lowers the gun and puts a comforting hand on John's shoulder. 'Don't worry, you'll be okay, remember what I said.'

She turns from John and inspects Harry's wound, then gives a viper stare to Hamish. 'Was this Major's idea?'

'Yeah, but it was for Rex. I find it hard to believe this shit.' He nods towards John, 'Sorry.'

Rena rolls her eyes, 'What's it gonna take? – What's that bastard up to.'

Hamish looks around the room looking to each, one by

one. He has made a decision to come clean. 'The starship Argon is fuelled up and ready to go. Now that the shield is perfected he's going himself, in person.'

Harry amazed, 'Himself... Major?'

'Yeah. Major always intended to be the first human to set foot on the planet Mars.'

'Dear God. Did Rex know this?'

'Yeah, he knew, Hal. That's what it's all been about, all for him. Money was only a means to an end.'

'But why?'

'He was a failed astronaut back in the pioneer days; he was deemed to have fucked up – grounded after just one space shot. He was seen to be an embarrassment to a failing space program. Then, after another disaster—a space-capsule burned in a ground-test. It was easy to include him in the casualty list... he was offered anonymity and a new identity. He took it, plus a platinum handshake.'

'But why, like this?'

'He never could get back on the team... with his know-how and contacts he was able to start the agency, but he couldn't get a ride... not with his background. The rest is history. He confided in me – Rex an' I were to accompany him. I still intend to do so.'

'But he's mad, you're mad,' says Rena. 'He used a nuclear engine, without regard.'

'That engine's okay, it was your uncle that fucked up, Harry. He used it ground to stratosphere and left a hundred year legacy of poison... part of Orkney is still under quarantine, under the cover of anthrax contamination. But oh no, we don't speak about that, do we? Do we?'

'Those were pioneer days,' says Harry, 'We were not so well-informed about–'

'Bull,shit. He knew exactly what was at stake, and he did it anyway.'

'Okay okay,' says Rena, stepping between them and pointing the gun again into Hamish's face, 'But you've got to be stopped. Major is going to pollute the planet before anybody gets there. Tell him, Harry.'

'She's right.'

Hamish casually brushes the gun aside, 'I don't think so.'

'Major's criminally insane. Tell him, Hal.'

'No he's not,' growls Hamish. 'Look, when we use the atomic engine it's only in deep space. Then, as Major says, it's like a gnat pissing in the ocean.'

'You mean the Argon has the same engine?'

'Exactly. Virtually the same as the original 1950's, Mandrake Experiment. It's hardly changed since the one sneaked out of Russia thirty years ago – some smart trader, your uncle. Those were the enlightened years: radar, television, lasers, jet engine, splitting the atom, hovercraft, nuclear power, computers, moon landings – all we've done in the last twenty years is develop, even the synthetics, just development on old technology – Absolutely nothing new.'

'My uncle was a patriot.'

'Yeah... so's Major. Anyways, all that time ago and we've hardly improved that engine... just ten times the size... not even that, just ten of the motherfuckers, stuck in a cluster, igniting in salvo. Whatever we say about the Nazis and the Ruskies, they sure were light years ahead of the rest of us. A crude slow-burn continuous atomic explosion... can't better that... Pow.' He makes a finger replica of a spacecraft and drives it across the front of himself.

'And you're still using it… in the Argon?'

'Yep. We're going to dock the Argon with Junairo3 and use its landing facility, then return to the Argon for the home trip.'

'And after that, what happens to Junairo3?'

'He never was going to bring Junairo3 back, Hal. God knows what's going to happen, I certainly don't.'

'But Harry's body is on Junairo3, you bastard.'

'I think Major intends to unplug it, him/ you.'

'He can't do that.' yells Harry.

'I think he can. He hates you, Harry. I mean he really hates you.'

'Bastard.'

Hamish shrugs. 'Shit, Hal, what can I say, I'm sorry. What a goddam mess.'

'We can't let him do it,' says Rena trying to make sense of it. 'When are you supposed to go back?'

Hamish considers the clash of loyalties. He makes a decision.

'Okay, we stop him. Major's already in transit – I'm to meet him in four hours on EarthlabOne, then…' he shrugs, 'Who knows?'

John, who has been quiet all this time, suddenly explodes into anger. 'What about me? Fock you, and fock Major. What about me?'

'Don't worry, John, I've got an angle.'

'Don't toy with me, Rena. Finish it now… Gimme a blue-nose and I'll do it myself… I'll eat the goddam thing.'

'Wait, John, just wait.'

'Wait? Wait, for what? I could kill all of you, I want to kill you all. It's all I can do to stop myself from ripping you all to pieces. Finish it now.'

Rena points the gun at John. She hesitates.

'He's got a point,' says Hamish, holding out his hand. 'Do it, Rena, before it's too late. Gimme the gun, I'll do it.'

Rena points the gun at Hamish. 'Back off.' She turns back to Harry. 'You an' this piece of shit make ready the SBS.' Hamish shrugs again.

Harry gives a worried look, 'What about you?'

'Don't worry Hal, I'll catch up with you. John and I have some history to catch up on... We got an angle. Right John?'

John shrugs, 'Whatever you say, Rena. As long as I get free of this shit.'

'No way, I'm not leaving without you,' says Harry, adamantly. 'Anyway, I'm injured, I need you.'

'Fuck it Hal, get to it. It's got to be this way.'

'But–'

'Please. I love you but GET TO IT. And get a dressing for your arm. I think I've got a plan. I love you, I'll always love you, no matter what.'

Harry gives her an odd look, 'And I love you 'no matter what' too, but–'

'And I love Jack fucking Daniels,' yaps Hamish, exasperated. 'Now, can we get moving, you're making a bloody mess on the floor?'

Harry and Hamish move off to the waiting SBS Orion.

CHAPTER SIX

Inside the shuttle, two attendant crewmen sit waiting. Harry pays them no heed. He takes a field dressing from the medic's case and turns to Hamish. 'You going to be mother and kiss it better?' Harry takes off his jacket, exposing his wound and the two syringes in his shoulder holster. Hamish sees the syringes... then rolls his eyes at the size of the wound. Harry awkwardly starts to dress the two-inch rut torn in his shoulder.

'Okay,' says Hamish, reluctantly, 'I'll do it. What have I got to do?'

'Sew the bugger up, for Christ's sake, what do you think, *spot*,bloody,*weld* the thing?'

'Okay smart-ass. You wanna shot with one of them...' he nods to the syringes, '... for the pain?'

'You kidding me? You just shot me full of novocaine.'

Hamish laughs, 'Oh yeah, so I did.' Harry gives a viper stare as Hamish makes a long, laborious hack job of extracting the bullet fragments and sewing the wound.

Harry glares in disapproval, 'Dear god, haven't you done yet? You do your own mending?'

'You don't like it, then finish the fucker yourself. I ain't no goddam nursemaid.'

Now Harry laughs, then instinctively winces as Hamish ties a knot and purposely roughly tugs the suture as he bites it off, then spits.

'Where the hell is she?' growls Hamish as he puts the dressings away, 'We got to go – You should never have left her.'

Harry looks through the open door to the building. 'What can she be doing?' As he speaks, Rena comes running from the building. 'Here she comes.'

As Rena runs she pulls on a flying suit, fiddling with the ties as she goes. A guard challenges her.

'Just a minute Ms–' she pushes roughly past him; he gives her a withering look, 'Have a nice day, Ms. Lansavitch, and kiss my ass.'

The guard walks back to the ward Rena has just left, to where John is held. The door is ajar. He gingerly pushes the door open. Inside the room, on the floor, is a widening pool of fluid.

Rena dashes along from the hanger to the runway and joins the slowly taxiing SBS. The craft's passenger door opens. She adjusts the last zip to her flying jacket – she is wet and naked underneath. She starts to climb in.

Half inside and hanging onto the hatch, she yells to Harry, 'Hit the button.'

Harry gives her a questioning look.

Back in the hospital ward, the guard pushes the door a little further. A pair of feet come into view, in the bubbling slime.

The Orion engines flame-up and the craft starts to move faster along the runway.

The ward door opens fully. The guard pulls back and gags, 'Jeeesus Chrisssst. Aruuugggh.' He vomits. As he does he pushes the door fully back, exposing the remains of John. The guard is compelled to look further into the room. Beneath the table lays the grisly remains of Rena's arm, pistol still in her hand, loops of decaying tentacles entwining the wrist like bracelets. The guard hits the wall-mounted alarm button. Klaxons blare out and more guards come dashing out of the main building.

Inside the shuttle, Harry leans from the open hatch and helps Rena inside. 'What happened, what did John tell you?'

'John didn't make it.'

One of the two crewmen receives a message over his headset. He looks back to the seated trio then to his crewmate and growls, 'Abort.' He then looks to his passengers, 'Sorry folks, we got a problem.'

As the engines start to die Rena is out of her seat in a flash and, in a single blow, knocks the crewman out of his console position. He lands unconscious at the feet of the second man. The man looks at Rena then back to his monitor. She glares menacingly to the crewman. 'Just touch the screen and say 'manual', I'll do the rest.' He does as asked. Rena sits in the empty seat and takes control. 'Hamish,' she calls over her shoulder, 'come up here and pick this guy up.' She beckons to the now semi-conscious pilot.

'And?' says Hamish as he complies.

'Dump them both out.'

'We're still taxiing, I can't.'

'Dump them out. Do it.'

Hamish manually opens the hatch. One crew-man wittingly slips onto the tarmac. He shuffles alongside the hatch and helps the other man. The hatch closes and the SBS starts to pick up speed.

Rena shouts to Harry, 'Purge and close off... Here we go.'

The two crewmen and the pursuing guards make for cover as the SBS booster engines fire. The craft suddenly surges forward. It leaves the ground immediately with an ear-splitting crackle. The craft is rammed almost vertical at an incredible velocity. After a few minutes two redundant boosters fall away, their descent violently interrupted by drag chutes. The Orion accelerates away to a thunderous crackle as it ploughs through the sound barrier. A secondary booster fires, sending it into the stratosphere, this now falls away and the craft hits the chemosphere, throwing out a widening gas trail. Hamish makes his way from the main cabin to another part of the craft, leaving Rena with Harry sitting by her side.

'Okay,' says Harry, 'What happened to John?'

'I told you, he didn't make it.'

'Explain that will you?'

'I blew his fucking brains out with a blue nose. Satisfied?'

Harry is shocked. 'Dear God, did he try to kill you?'

'No.'

'You killed him in cold blood? You executed him... Dear God.'

'It wasn't like that.'

'Dear God.'

'Will you stop fucking saying that,' she snaps. 'For Christ's sake say something else...' she calms slightly. '... we got a problem, remember?'

'Did he talk?' Harry stares at her, she doesn't look back or answer. 'Did he say...?' His stare intensifies, boring his eyes into hers... again nothing. 'Did he damn well say what got rid of them? What about the children? Christ almighty, Rena... answer.'

'Okay okay. I know.'

'You know. You know what? The children...?'

'I know it all. You're going to love it.'

'He told you how we can get rid of them?'

'That's what I said, didn't I?'

'Bloody hell, what's got up your nose?'
'How many friends have you killed today, Harry?'
'Yeah, sorry.'

Her eyes flood with tears. 'I'm sorry too, Hal. I had to do it, I promised him.' She turns and kisses the side of Harry's face. He turns and kisses her mouth.

Hamish re-enters and sees them kissing. 'So, you two going to dock or fuck…' Harry, follows his directing gaze through the porthole to EarthlabOne, looming toward them… Hamish looks to Rena, '… or maybe just go on smooching and fly past the goddam thing?'

'Don't worry, I've got it covered.'
'I don't do worry, Lady.'

Alongside EarthlabOne, inert and tranquil, lies the great starship *Argon*: It is half the length of its predecessor, Junairo3, but twice its girth. A shapeless canister with its entire propulsion unit contained within.

Rena turns to the monitor screens. A sudden crackle of static, then a voice:

'And where the hell you think you're going?' Major's face leers from the monitor screen.

'Let us in, Major, or I'll ram the Argon. I mean it, I'll hit the causeway and set the Argon adrift.'

Harry, alarmed at Rena's suggestion, grabs her arm. 'Hold it, Rena.'

Rena pulls away violently and turns on Harry.

'Keep out of this or I'll fucking kill you,' she screams at him. Harry recoils in shock. Rena glares back at him… A long, nervous pause.

'Hey, that's not EarthlabOne,' interrupts Hamish, looking away from the monitor he yells out in surprise, that's the flight deck of Argon. He's onboard already.'

'Smart, very observant,' says Major from the screen. 'You've blown it, Son, I'm going solo. By the way, I've already started the countdown so I suggest you back off out of range – Over and out.' The screen goes dead.

Rena considers for a moment. 'If the Argon has the atomic reaction engine, half of the bulk of the ship must be radiation absorbent ballast, right?'

'Yeah. So what?' says Hamish, puzzled.

'So what? So I'm going to do what I said, I'm going to ram.'

Hamish, shocked at the suicidal suggestion, grabs at the controls. 'You crazy bitch, over my dead bod–' before the word is finished Hamish is hurled across the cabin, hitting the gunwale with a sickening thud.

Rena turns back to the controls and switches a button. The craft immediately banks violently. Harry is now standing behind her. He has one of his syringes held behind his back.

Without turning, Rena speaks. 'Don't do anything stupid, Harry. John wouldn't spill. In spite of everything he said, I guess he was loyal to the last, to his own kind, *and* to us. He couldn't pick sides, so I offered him a deal.'

'A deal, Rena? This isn't a game of tag for Christ sake, there's no way back.'

'It was the only way.'

'No. God no.'

'Put the syringe away Harry, you won't need it.'

'I won't need it?'

'No, you won't need it. Rose was weak and confused, John was, too. He was a hero but he wasn't strong enough. I'm as hard as nails.'

'What are you getting at?'

'Them mothers are going to pay for my father, Harry.

You want to know my father's real name?' Harry shrugs indifference, 'It was Crabbe.'

'Crabbe? ... *Crabbe*. Not... ?'

'Yes, Buster, the same: Commander Lionel Crabbe, HRN.'

'Dear God.' gasps Harry.

'My father was in at the beginning with your uncle, Lord Melrose. Your uncle used the NPP, *nuclear pulse propulsion* engine invented by a Russian, Stanislaw Ulam, in Nazi Germany. With my father's submarine know-how, your uncle put a shot to Mars. My father had the authority to commission a new submersible unit, an addition to the latest post-war submarine development. Your uncle was, shall we say, the go-between... Mr. Interlocutor, so to speak. What they sent into space was virtually a submarine.' She shuts her eyes and an image appears: Orkney, the 1950's rocket-launch in progress. A huge iron-clad craft of enormous weight is lifted into the air by a continuous atomic explosion, ramming it into the heavens and laying a vast, poisonous shroud in its wake.

'But the payload, Rena, a submarine for god's sake, it must have been enormous.'

She opens her eyes and the image is gone. 'The engine was that goddam powerful, Hal, the size of pay-load was never a problem.'

'How come this information wasn't in his papers?'

'That's the way they wanted it. Lord Melrose had the power of veto. After the Korean War there was a lot of stick flying around.'

'Stick?'

'Recriminations, Hal... the new Rolls-Royce Nene, turbo-jet engine technology your government traded for the Russian/Nazi nuclear pulse-engine – the jet engine was used in the Mig fighter. America was, to say the least, pissed. ... Your uncle respected my father's wishes.'

'But–'

'Just listen, Hal, not much more: So, that's why Rose needed you in. You had the access to your uncle's papers, you were too dangerous on the outside.'

'Why didn't she just kill me after she got what she wanted?'

'She should have, but... she fell in love with you, Hal. The gods do love... Boy, do we love.'

'What about Barney?'

'He's all human, Harry.'

'How can you be certain?'

'Take my word, as one that knows.'

Rena stares at him, Harry does not push the point, 'Carry on... your father...?'

'So, after the British space shot slit the bag, my father had no choice but to defect to Russia... he was scapegoat for the Korean/ Mig15 debacle... that was the deal. I know what you're thinking, Hal,'

'Did the... thing get to him?'

'Yeah, they got to him, the entity... the brood. He was one of them... one of us. That wasn't his body they buried in Chichester Cemetery in the 50's, my father died in space.' She turns and looks at him. Harry holds the syringe out in front of him.

'You know what I have to do?'

'Just listen, Harry. Give me a minute.' Harry lowers the syringe. 'My father would have been over eighty-years-old when he met Major... he looked forty. That's part of it... we're gods, remember.'

'Did Major know?'

'Hell no, of course not. But he did know my father was

a Russian plant. He had no choice. He took my father and he got the engine and the USSR's Mars portfolio... that was the deal. After the first Junairo, after Rose, we lost almost two years–'

'You and me, both. I spent that time in the bloody slammer. Major still owes–'

'For god's sake, Harry, will you fucking listen?'

Harry lifts the syringe again. Rena calms slightly and continues. 'Sorry. After Junairo1, Major had to gain time, he couldn't wait for a new window, he needed power. The Russia House was, as Major put it, 'light in the kitty'. So instead of money, they covered the pot with the NPP nuclear-pulse propulsion unit, designs for, that is – my father came with the deal: Buster Crabbe was alive and well and living in Idaho... as a sleeper.'

'And what happened to my uncle, did they kill him?'

'I'm not sure. The story is he blew himself up developing the engine. I do memories, Hal, not ESP.'

'Didn't your father say?'

'As I say, he wasn't sure... with Russia, in the Stalin days, nothing was certain. Take the death of Hitler... we still don't know for sure.'

Harry covers his eyes with his hand. 'Dear God.'

'You're saying that again, Harry.'

'Sor/ carry on.'

'Anyway, my father had his own ideas. He was going on that trip for one reason only... the same reason he defected to the Russians... to use that engine to kill the whole goddam brood. I'm going to use his plan.'

'So, you knew about your father all along and you let me–'

'No I didn't. Damn you, Hal, I didn't. I only found out a month ago.'

'I'm sorry.'

'You're saying 'sorry' a lot, too. You don't want to piss me off... not today, not now.'

'S... so, carry on.'

'All that time, I didn't know. And I'm sure, to her dying day, my mother didn't know.'

'I really am, dare I say, sincerely sorry.'

'He left a letter for me – if he failed, he wanted me to know.'

'But if he was one of them...?'

'Somehow it knew what he was about. I saw them all die, Hal. I saw the video. The entity got into the ship and blew every one apart. It was horrible.'

'Are you... were you–'

'Before you ask, he was my real father. I was a normal kid right up until... until John.'

'Did it hurt, you know...?'

'Harry, I don't have time for this. Put the syringe away, I can handle it.' At the controls, Rena makes another violent manoeuvre.

Harry's thoughts now concentrate on the immediate danger. 'If you ram him, you're kill us all. We'll never survive.'

'Show some grit, Harry. The ballast is probably light mica-shale or pumice-concrete, whatever. It's soft, not as absorbent as lead, it has no density.

'Probably? —Suppose it *is* lead?'

'How in hell do you think they'd get that much lead up into space? You might just as well say gold. Christ, I wish you'd think before you speak.'

'Okay, okay. But what about Hamish?'

'To hell with him. He didn't give a damn about you. He shot you... twice, and he shot to kill, didn't he? Let him take his chances.' She calms slightly. 'Right, here we go. Take my count-down from ten. We walk the cord on zero. – Ten ...'

Harry closes his eyes in resignation and puts the syringe away.

'Nine... Eight... Seven...'

The Argon starts to manoeuvre away from its mooring – the SBS Orion is closing fast.

Harry's eyes shut tight as Rena's voice booms out.

'Six ...'

The Argon's huge power stack leaps into life with an eruption of vacuous gasses. But it is too late. The SBS is on a collision course.

'Five... Four ...' The Orion is about to hit. 'Three...' Harry's eyes widen in terror. 'Two... One... WALK.' The SBS closes on the starship. 'Zero.'

—Collision.

CHAPTER SEVEN

Harry stands alone in the gloomy mist of the abyss. He strains his eyes to see, but sees nothing. He calls into the blackness, 'Rena, can you hear me? ... If you can't speak just walk toward my voice,' he holds his hand out in front of him, 'Just take my hand, Rena. Take my hand. Please, Rena, answer.'

The white wolf momentarily appears, superimposed over Harry, and then it is gone. Harry again scans the bleak landscape, but it's too dark and misty. He seems to sense a presence, and moves slowly off through the gloom into its direction. A hand stretches from the swirling void. Harry grabs it in the darkness.

'Rena, thank God. I thought for a moment it was Redman... Ahgggggg.' Harry gives a piercing high-pitched scream as the blade of an axe, flat side, flashes across his forehead.

Redman's gargoyle face looms out of the void, 'What say, Harry? Bloody bingo. —Nice to know you're *thinking* of me.'

The circle eye in the centre of his forehead is still smashed and weeping black bloody slime. The cowl of skin is still ripped open and exposing his lidless eyes and gash mouth. Harry backs off in terror, falling backward into the snow. The axe slashes. Harry crosses his arms and catches the shaft in the 'V' of his wrists, stopping the blade just short of his face. He instinctively kicks up a knee into Redman's withered groin sending him tumbling, landing in a heap in the snow.

Redman flaps his arms and tries to right himself onto his half-formed legs, screaming and spitting insults. 'How's Rosette, laddie, seen anything of her lately? I like your son, Hal. He looks like Rose, nothing like you, you fucking fruit. Haaa. Christ, Harry, I can't get up. Give us a hand, old boy... the legs go first, so they say. How's Rena treating you? I heard she went down on old Johnny boy, haaaa ha ha.' Redman manages to stagger up, propping himself on the shaft of the axe. 'Hey, laddie, how'd you like a big wet kiss?' He tips his head back and blows out a mouthful of blooded mucus. The snatching tongues whip across the short distance, looping Harry's arms, dragging him through the snow towards the gaping mouth. As Harry is pulled the last few inches, he manages to free one hand and grab the handle of the axe that Redman is using for support. He pulls it upwards, forcing it to jerk free. It flies up between their faces, severing the scarlet tentacles in a burst of blood. The recoil throws Redman into the snow. Harry whips out one of the syringes, but before he gets a chance to use it Redman hurls himself back, grabbing Harry's arms. The shortened tongues leap again, smearing Harry's face with bloody gore. With all his strength, Harry head-butts Redman, slackening his hold long enough for him to ram the needle through the underside of Redman's chin, up through the tongues, through the roof of his mouth and into his brain. The plunger deposits a full barrel. Redman recoils, pulls back and covers his contorted gargoyle face with his hands. Harry sinks to his knees and leans back on his haunches and waits.

Redman wails, crying out in agony, 'It's burning... It's killing me... I'm going to die ... fucking laughing. Ha ha ha, he he.' He pulls out the syringe and takes his hands away from his face. He is weeping with laughter, 'Here,' he throws back the syringe, 'You'd better have this back, you'll need it.'

'Dear God.' gasps Harry.

'There's no God here, laddie, not, this time, old luv, old sport. Nothing works here, not in fucking Hell. Ha ha ha hoooo.' He stops laughing and adopts a serious posture. 'Sorry Harry, have to dash. Here comes your Gorgon piece of pussy—make her give you head... know what I'm saying? Haaa, he he.' He tips his head back: a dozen fiery-blooded tongue-stubs jaggle lewdly in his mouth, then he is gone.

'Harry.' Rena's voice calls out from the blackness. She steps into the light. 'Hal, thank God, you okay?' he doesn't answer. 'Harry. Hey, answer for Christ' sake.'

'Did you see him?' says Harry at length, 'Did you see Redman?'

'Redman? No, I didn't. Didn't you kill him?'

'Yes, once. No one stays dead for long in this place – I must have thought about him. I shoved this up his throat and all he did was laugh. It doesn't bloodywell work here.'

Rena gives a chuckle. 'You shouldn't have told me that, Harry... me of all people.'

Harry gulps in a deep breath as he realises what he has just revealed. Rena sees the fear in his eyes.'

'Don't worry, no one's ganna hurt you, Honey, just ganna fuck you a little.'

She waits a moment, 'Well?'

Harry, hardly understanding, manages to find his voice. 'Well, what?'

'No one's going to hurt you, Honey, just ganna fuck

you a little'... What's it from?'

Harry falls in. 'Oh, 'Last Exit to Brooklyn,' almost – right?'

'Right on. —You okay?'

'I'm fine, now. How do you feel?'

'It's okay... I still have the memories, but nothing else – I don't want to hurt you, Honey, just...' she raises her eyebrows and smiles. Harry gives her a nonplussed scowl. Rena shrugs and continues, 'God Harry, the things I know.'

'Did we hit? You stayed after zero?'

'We hit. The Argon's shell pierced, we're jammed into its belly...' she smiles, '... it wasn't lead. Our hull was holed but the SAMs were on the job when I walked.'

'Hamish?'

'Hamish didn't make it, Hal... a shard of metal tore his suit... sorry.'

'Damn.'

'We can't get to Major until he closes with the Mars satellite, Phobos. So, from now on we don't have any contact with him, not a single word.'

'Why not?'

'Because when he looks into the screen, after the SAMs power up our ship, he'll see the mess from Hamish splattered all over the cabin.'

'And...?'

'And... he'll assume we're all dead.'

'Which begs the question,' says Harry, 'What about us... our bodies?'

'God' sake. We're alive, ain't we? I instructed the SAMs to suit up your synthetic and my body and put them on the suit-rack. We look like we're empty suits... I hope. When he gets within gravity of Phobos, we're going to pull out and race for the Junairo3.'

'And that's your plan?'

'Yep. That's my plan.'

'And how the hell are you going to get the Orion out?' says Harry, thinking Rena has missed the main point, 'you said it's jammed.'

'Fire the retros.'

Whaaat? Are you're crazy? You'll blow our bodies to bits.'

'Yeah, trust you to find the weak spot. —Anyway, as I told you, we're suited up.'

'Oh, well, that's alright then, that'll save us... I'll tell that to Hamish, shall I?'

'Listen. If Major gets to Junairo3 first he'll pull your plug. You do realise that?' Harry is silent, after a few moments she continues. 'You've got time to consider. My idea is for you to stay here, I locate on my own... I don't trust myself in that body, I'm still frightened of what I'll do.'

'No, way. You're not going back on your own. You're going to blow it up on purpose... that is your bloody plan, isn't it?'

'And them, Harry, HOW DOES THAT GET RID OF THEM?' she screams into his face, 'I wish you'd fucking think before you talk, Harry.'

'And I wish you wouldn't swear.'

'*What?*... You're about to be blown to kingdom come, I'm a homicidal/ suicidal maniacal alien entity, and you wish I wouldn't swear. God, I love you, Harry, I fu – frigging love you.'

Harry is unmoved by her friendly mood-change. 'So, what is the plan, Rena... how was it done before? Rose said they were here as early as the Palaeolithic period, she said Cro-Magnon man was their/ your legacy. You said I'd love it?'

'Okay, we still got time. She was right in a way: We/ they were what you might call the Missing Link. We came the same way, meteorites. We cloned and fed off Neanderthal man. We took only what was needed, their bodies, their blood and their intelligence. That was their/ our downfall, their intelligence. Neanderthal man did not possess the spark for change, not like that of Cro-Magnon man. We died out with the Neanderthal, leaving the road clear for the Cro-Magnon... that's what Rose meant. The second time, Mesopotamia – meteorites again – this time they/ we destroyed ourselves, in two separate ways. You got one: they fraternized with the locals and picked up a very nasty habit... nectar.'

'Nectar?'

'Yeah, the drink of the gods, Harry, and it acted like a hallucinogenic... the natives took it to stave off hunger and cold. It was basically honey, distilled wine and, here's the bit, Harry... leaves.'

'Leaves?'

'Right, LEAVES,' she yells into his face. 'So this is your new way to piss me off, is it? You're going to repeat every goddam word I say?'

'So... No.'

She gives him a reassuring smile. 'In the translation, it says Gilgamesh searched for it under the sea. Not under, over. Over the sea, the early traders brought it. The locals chewed it... coca leaves, cocaine.'

'Oh, my, God.'

'Yes. And when they took it, it separated the beast from the human element. It allowed them to hallucinate. Hence the amazing stories, the Titans.'

'And...?'

'And they got hooked on it. Now *novocaine* is a different bag of worms. Because it is refined and synthetically concentrated, it doesn't just separate... it blows the two elements apart, literally.'

'Okay, that was one way. What's the other?'

'Sex, they liked it. When they were high on nectar, it made them horny as hell.'

'Horny? ... Oh, I see, aroused.' He gives a disapproving, distasteful shrug.

'Yes, 'horny.' Hence, killing the beast–' She stops and studies him. 'What's the matter Harry, you don't like the term, 'horny'? Don't tell me you're embarrassed? God. You English are so prissy.'

Harry gives a wearisome, 'Go on.'

'Yeah, *aroused*. It's in the bible, Genesis – what John told us – 'giants in the land laying with the daughters of man,' Whooo,wee. Did they like a lay. Sex and drugs and rock'n'roll. What's new?'

'So, how did that get rid of them?'

'Sex was the beginning of the end, they didn't like their way no more... your/ our wicked ways corrupted them... sex, drugs and whatever. We/ they just gradually died out. There were a few strongholds. The last of us died out in the late sixteenth century, in Transylvania.' She laughs and makes a cross with her fingers, 'Don't worry, Harry, it's not hereditary, Barney's okay... believe me.'

'You keep using the terms 'your', 'us' and 'we'.'

'Damn you, Harry, have I laid so much as a finger on you? Them, them!' She breaks down and weeps. Harry is shocked, he's never seen Rena lose control before.

'Sorry,' he hugs her.

She pulls away angrily and yells back at him. 'Christ

sake, will you stop saying 'sorry.'

'Okay okay. Jesus. Go on, go on… your plan.'

'Once I blast free I'm taking the SBS and I'm going to beat that bastard back to Junairo3, then… Do you really wanna know my plan?'

'I think I already know it.'

'I'm going to orbit Mars in the same trajectory as the Enkidu belt and lay a path of radiation and destroy the whole goddam motherfucking brood – please, forgive my French.'

'And then?'

'And then… I'm going to pull the rods out and leave it in orbit for eternity.' She calms slightly. 'You'll be okay, Harry, it's the only way now. We got nothing together. On Earth I'll always be thinking of killing you. Out here…' she looks out over the bleak terrain, '… Christ, in this hell-hole I'd always be thinking of killing my goddam self. My way is the only way.'

CHAPTER EIGHT

On the flight deck of the Argon, Major sits looking into a bank of monitors, one of which displays the planet Mars. He speaks out into the onboard mike, 'Give ETA of Mars orbit, and rendezvous with Junairo3?'

A generated voice answers: 'First option. Argos to orbit Mars is in thirty-five minutes, copy monitor two for precise details.'

Major turns to the second monitor, which is giving out digital information; trajectories, landing site options, etc., and continually updating. Major speaks again, 'Give projectile damage... upgrade every five minutes. Give damage report, on monitor four.'

Immediately a fourth monitor begins to give information; a visual rundown on the ship's damage. Halfway through, the computer-generated voice breaks in:

'There is activity in damaged ballast section. The appended SBS computer has been assigned to manual override.'

Major looks perturbed. 'Can you intercept? It's priority.'

'Negative.'

'Try.'

'Negative. Incoming communication on screen three.'

Rena's face appears on the screen. 'Major, are you ready? Last one home's a faggot.' Her face disappears before he can answer. As the screen goes blank a violent tremor shakes the ship.

The SBS, breached into the side of the huge spacecraft, fires its retro rockets. Gasses void into the vacuum like clouds of degrading steam. Again and again, they fire, burn, and shut down. Gradually the SBS begins to tear loose. A final roar and the shuttle is free. The Argon veers violently off-course under the recoil.

The SBS is sent spinning wildly, but its gas-jets quickly stabilize and sets it on course for Mars' moon, Phobos. The starship Argon has drifted aimlessly off course, its starboard jets pulling it slowly back under control. Now fully stabilized the massive atomic engine fires.

Inside the shuttle, Rena throws a switch and Major's face appears on a monitor screen.

'I win, Major,' she growls into the mike, 'I dock in two minutes, your ETA is ten. You got the power but we got maneuverability. I suggest you overshoot, round Mars and go home.'

'That is not an option,' says Major, grimly, 'I have to dock with your SBS. The Argon is leaking like a stuck pig. Your final blast knocked a hole in the ballast hull. The cabin is showing three times danger level. I'm contaminated. I have to dock. If you say no I'll detonate the power-stack and waste us all...' he is silent for a moment, his solemn face stares out at Rena, '...that, or I have a deal.'

'I'm listening,' says Rena.

'Let me onboard. We rendezvous with Junairo3 and I'll take its lander... I don't need access to Junairo3, just its lander. I have to set foot and raise the flag. I'll just have time.'

'And that's the deal?'

'That's the only deal. It's all there is, Rena. Or I waste the lot.'

Rena considers. 'Okay. I'll fire a lanyard and you come aboard. You give over command of Argon and when we get to Junairo3, you go straight to the lander, then we'll take you to orbit. That's my part of the deal... I was going that way, anyway. Take it or leave it. I need to know, now.'

—In the Badlands, Harry stands looking out over the abyss. His eyes scan the swirling vapour. He seems to recognise something, he smiles then turns to the silver beam of light stretching out to infinity.

On the SBS, Rena fires the gas-jets, closes, and docks with Junairo3. She goes straight to the airlock, enters and closes off. Once onboard the starship she walks on to the flight deck and begins the recovery procedure. As the console comes to life, Harry's spectre manifests through the decking of the shuttle. Rena, busy restoring control does not see him as he locates with the synthetic. Grey Harry removes the pressure suit then takes a syringe from the medical cabinet. He enters the airlock through to Junario3 and stands unnoticed behind Rena, waiting. Major can be seen through the porthole, tethered on his umbilical, stretching out from the Argon to Junairo3 alongside the docked shuttle.

Janairo's airlock opens and closes then floods with gas. The inner doors open with a hiss and the hatch flings back. Major stands hunched and covered in dust and dry ice. He unclips his black-domed helmet and lifts the visor. He is bleeding from the nose and ears. He speaks in faltering voice.

'I'm going to move straight to the lander... I'm deadly. Wash yourself when I'm gone. Whatever you think, it was all for America.'

'Bullshit,' she yells, angrily into his face, 'You stay put. You're going nowhere. Everything was expendable except the agency, you said... Bull,shit. It was your goddam ego, it was all for you. You never got over the fact that you fucked up as an astronaut. This was to make amends. Well, sor,ry... Not on this tide, Buster. ... Recognise that name?'

Major moves toward Rena. He forces a smile as he speaks, 'If you don't move soon you'll be contaminated. I don't give a damn. I'll waste you and still take the lander. I won't need you to take me there. There'll be enough fuel, It's one-way remember.'

'You're not going, Major. It's all been for nothing. You'll waste me? I don't think so.'

Major stops dead in his tracks as he sees the slime drool from her widening mouth. The grotesque gargoyle head tips back and the tongues leap, a dozen fiery red tentacles leap out at terror-stricken Major, tipping him against the decking. Rena moves towards him for the kill. She is stopped short at the end of a razor sharp hypodermic needlepoint at her throat. Harry stands behind her, one arm around her waist.

'Let him go.'

Rena reacts, Harry pushes the needle a little way into her skin.

She retracts the tongues and speaks. 'No, way. Let me do it, Hal... just let me. I'll finish us both. It's my only way out. You can take Junairo3 back to Earth, with luck your body will mend. I'll take the Argon and waste the brood. I won't even have to pull its rods, it's leaking anyway. Right, Major.'

Major nods, yes.

Harry steps closer to Rena, the needle still in her throat. A long, nervous pause, after which Harry breaks the silence. *'Nobody move – I've got an idea.'* He gestures to Rena. She looks back at him, over her shoulder and frowns, not understanding. He gestures again.

Rena falls in. 'Jesus Christ, Harry... *The Italian Job* – Are you fucking kidding me?'

'No, just checking. Just making sure there's enough of the old Rena left in you... I'm not spending eternity with a bloody bore. And I do have an idea. And we do have a chance... slim, but a chance.'

Major staggers up, Rena makes a move. Harry strengthens his hold on the syringe. Rena backs down.

'Some, chance. You know I'll kill you the first time I weaken – I can't trust myself. My way's best.'

She moves again, Harry jabs the needle in deep, Rena jerks to a halt. She watches the clear, yellow fluid move into the needle.

'Please baby, don't make me do it. Let him go... think of us. We do have a chance.' He turns to Major. 'Get into the lander, Major, quickly.'

Rena struggles, the needle goes in deeper. Major starts to move out.

As he disappears through the lander's hatch he gives a parting word over his shoulder to Harry. 'Thanks, Mandrake.'

'Three years and you finally called me, Mandrake. —I never thought I'd hear it.'

'Yeah, well... don't go expecting this special treatment all the time – see you in the next life.'

'Dear God, I hope not.'

As the hatch closes, Rena screams at Harry. 'You let the bastard go. Christ, Harry, he tried his best to kill you.'

'Trust me, Rena. I promise you we've got a chance. Move to the coffers.'

He leads her slowly across the cabin to the bank of coffers and indicates to the one next to the casket that holds his own battered, comatosed body. With one hand carefully holding the syringe to her neck he lifts the lid of the empty coffer.

'You got to do this very slowly, Rena. You've got to get in. Don't make any sudden moves.'

Rena starts to climb in. 'It won't work, Hal. Just let me take the Argon and blast a radiation belt. Once I've done that, I'll die happy and–'

'Just get in. We'll do it all, and we'll still have a chance.'

Rena gets in the coffer. Harry starts to pull the crystal lid down and slowly begins to pull the syringe from Rena's neck.

'You stupid jerk, I can easily smash this. The minute the needle comes out I'll have your hand.'

'For Christ's sake listen, Rena, just listen.' The coffer is now half closed. In a burst of anger, Harry jerks the needle full out of Rena's neck and squirts the liquid into her face. While she is distracted he slams the lid shut. Shocked, Rena clasps her hands protectively to her wet face anticipating the burning... Nothing. She takes her hands from her face and looks at them, and then sniffs them.

'What the hell was that?'

'You didn't think I'd really do it, did you, it was just water?'

'I don't believe you. I saw it, it, it was yellow. Where would you get yellow water from, up here?'

'Where do you think?' he raises an eyebrow, 'Now shut

up.'

She considers, then suddenly realises. Shocked and horrified she holds her hands in front of her face in disgust, and inspects. 'You bastard.' she yells as Harry walks off. 'You didn't, did you, Harry? Please say you didn't?... You peed in my face... dog's pee, you bastard. You did, didn't you?'

'You'll never know. Just lie there. I'll plot Junario3... Get ready to walk.'

'What, and live in Hell? No way.'

'No, it's not hell. There's green countryside there, I'm sure. When I first walked, all those years ago, I'm sure I saw it. I'm sure I can find it again, Rena, I promise. And when Junairo3 fails we both go together, side-by-side.'

'And the, Enkidu Belt? The brood?'

'I said we'd do it all... Major will do it, with the Argos.'

'Are you kidding me? He's got what he wants, he won't do it.'

'He'll do it... I have an offer he won't refuse.'

'He's dying. What can you offer a dying man?'

'Immortality.' Rena is silent. Harry places his hand on the screen and speaks to the open mike. 'Abort landing programme.'

Immediately a redundant screen lights up with Major's face, he speaks with pained voice. 'You've aborted, Harry. What's the problem? Make it quick, I'm bleeding badly.'

'You've got to take the Argon with you. Because of the damage, we can't direct it, you have to do it manually. Dock and set it on course for Mars orbit, the Enkidu belt, and let it empty its atomic fuel. You don't even have to pull the rods.'

'Harry, I'm dying... I don't got time for this shit.'

'This is atonement time, Major... time to give a little back. You can wipe them out. Just think of the glory, your message back to Earth, sacrificing your return trip and your life to rid mankind of the beast.' Harry quotes Major: 'Nobel Prize, at least.' —I remember your saying that to me, a lifetime ago ... posthumous, of course. All those people who had you down as a coward. They said you lost your nerve when the going got tough. You won't pass this up.'

Major is hooked. 'Okay, I'll dock with Argon and plot the course, but I won't leave the lander, Hal. I don't think I could if I tried.'

'No need.' says Harry.

'I never opened that escape hatch, Hal... I never blew those bolts... All those years ago they never forgave... I never opened it, just a malfunction... they must have done it... a slip-up at control... everybody was covering ass. D'you think history will believe me after this?'

Harry just nods, Major forces a smile and continues, 'Okay, once I've docked I'll set coordinates, and that's it... that's for America.'

'It'll do, Major.'

The side panel blasts away from Junairo3 and tumbles off into space. It reveals the lander, a small, awkward-looking craft huddled in the exposed bay. A gas-jet eases it from the mother ship and pushes it on toward the Argon. It fires its lanyard and pulls itself to dock.

Harry stands looking expectantly into a blank screen.

After a moment the screen flickers to life with Major's face. He speaks with difficulty, 'Okay, I think I'm just going to make it... See you in hell – Over.'

'Not if I see you first – Over and out.'

The lander docks with the damaged hulk of the starship Argon. After a few moments, it manoeuvres away. Once clear, Major fires the Argon ballistic engines, sending it on course for Mars orbit.

Junairo3 fires its atomic-reaction engine, ramming the sleek craft into oblivion, blowing out its massive veil slipstream.

Harry and Rena's gossamer images manifest into a blood-red landscape of Mars. Their eyes are attracted to the scarlet sky, to a tiny glow that grows ever larger as they watch. It is the Junairo3 lander. Drag chutes open and its ballistic retros fire: A perfect landing. Harry and Rena are, in spirit, audience to mans' first physical footstep on Mars. The lander hatch opens and Major steps out and collapses to his knees onto the bleak, red surface. He bends forward and, through the glass dome helmet, makes to kiss the dusty ground. Rivulets of blood run down the inside of his visor. He rocks back on his haunches and speaks with extreme difficulty into his helmet mike:

'I claim this territory in the name of the United States of America.' He takes a small folded Star-spangled Banner and places it into the crusty surface, then tips forward, face down onto his helmet, now opaque with blood and vomit. The prostate figure moves no more. The fine mist emission from his life support gradually fades to nothing.

CHAPTER NINE

They stand for a while looking down at Major. Harry gives a glib salute.

A cloud of mist encircles them, from which the silver cord of light streaks to infinity. Harry takes Rena's hand and they walk into the void. He speaks from the swirling vapour.

'How do you feel?'

'Okay, I guess,' answers Rena, 'I still got the memories but not the urge to kill. How about you?'

'I'm fine, glad to be rid of that grey hulk. The shoulder was beginning to give me gyp... novocaine was wearing off. Nice to have my real body back... what you dream is what you get.'

'No regrets, Harry?'

'None... except I wish I could have kissed Barney goodbye.' He considers a moment, 'Do you think I'll ever see him again?'

'You'll see him, Harry, every night of his life. He'll come to us in his dreams... We'll share his dreams.'

'But he'll never know, Rena. He'll never really know, I mean properly. Will I ever see him... proper?'

'Don't think about it too much, Hal. We'll figure a way. Maybe we can teach him to walk the silver cord.'

'No, not that, I want him to be just a normal boy.'

'Anyway, who knows – we got time.'

The void is now a flare of monochrome yellow. It

gradually sharpens into a field of golden daffodils set against an azure sky. Harry and Rena are sitting in a field of flowers swaying timelessly in the breeze. Harry is bronzed flesh-and-blood, Rena, beside him dwarfing him once more – they embrace. Harry lays his head in her lap.

She touches his face with a blade of grass, 'You bored, Harry?'

He considers and looks up at her. 'Are you?' Rena doesn't answer. Harry looks yonder across the fields to where the mist is just visible, 'Just look into the mist, Rena, the bits... think of their misery. I mean to help them... I don't know how yet, but it'll come. —I'm sure the old man, my uncle, is out there. I'll find him.'

'Sure you will, Hal. —Okay... where are we up to?'

'Oh, we just finished the Trojan Wars. And no, I'm not bored.'

She bends over him and kisses the side of his face. 'If you are bored, I could always conjure up Redman for a bit of excitement.'

'You kidding, don't even think about it—I mean, don't think about it. Honest, I'm not bored. Christ, Faustus gave his soul for this.' He considers a moment. 'Do you think we've lost our souls?'

Rena now considers. After a few moments she answers, 'Naaa... we done nothing wrong, Harry.'

'Oh? Genocide is nothing wrong? We're all God's creatures.'

She considers again, 'Naaa.'

Harry lays back in deep thought. Suddenly, his eyes flash open to a subliminal image of Redman. The grisly severed half tongues jaggling lewdly in his gash mouth. He is still drooling blood and slime. The smashed Cyclops eye weeping black ooze, the torn cowl skin revealing blooded lidless eyes. The spectre takes a stumbling step forward. Harry just waves it away. Redman evaporates.

Rena puts her hands on Harry's shoulder and pulls him back to lie on her lap once more. 'You okay, you were miles away?'

Harry smiles. She bends to kiss his mouth.

He puts his fingers over her mouth and she kisses his hand, 'I bags no Frenchies,' says he smiling, 'Right?'

She punches him playfully in the arm then takes his hand from her mouth and kisses him, a big wide-open sensual kiss.

CHAPTER TEN

The monochrome yellow gradually turns to a vivid blue, a surreal field of cornflowers surrounds a beautiful classic temple of white marble. An immeasurable period of time has passed.

Harry and Rena are sitting in the same position as before, the temple in the foreground. He rests his head in Rena's lap then turns his ear to her stomach.

'Can you hear it kicking, Hal?'

Harry listens, laying his hand on Rena's swollen stomach. He smiles and remains for some a few moments listening, then lifts his head. 'So... whose child do you think it is, then?'

'WHAT? What the hell do you mean?' She turns on him outraged, slapping him around the head. 'You think I've been sleeping around... you calling me a tramp? Who's could it be? Haymish's? Redman's? Fuck you, Harry.' Laughing, Harry protects himself with his hands as Rena continues. 'Major's, for Christ's sake? goddamit, Harry, you got some fucking neck.'

'I didn't mean that, honest. I meant 'whose'... pink or grey: English gentleman or Wolfman, which one of me?'

She stops hitting him, and considers. 'Oh... my... God. I hadn't even considered... Pink. Hold on...' she does a mental calculation, '... No, grey, maybe... No no, pink. Definitely pink. Sorry for swearing'

'You're sure now?'

'Yes, I'm sure... totally.' After a few moments, she smiles. 'Shall we call it, Rover, Harry?'

Harry, nonplussed, 'I thought you said you were sure?'

'I am. – *'Shall we call it Rover, Harry'* – what's it from?'

'Oh... *The Kidnappers*, Philip Leacock, circa nineteen-fifties... I think.' He smiles smugly and leans his head back on her stomach.

Rena gives an, 'I win' smile. 'So, what happens when she's born – it's going to be a girl, I know.'

'Don't worry... we'll figure something out, we're still on a learning curve remember.'

Harry listens to the baby intently. After a few moments, he sits up, his happy expression changing to one of surprise, then to horror. He turns to Rena with a frightful look on his face. 'Oh my God,' says he, alarmed, indicating to her stomach, 'it's full of stars.'

Rena gives him a questioning look. 'Give me a break for Christ sake, Harry –*Two-Thousand-and-One: A space odyssey*... Kubrick, circa nineteen-seventy-eight.'

Harry smiles, 'Bloody bingo.'

Inside Junaro3, all is calm, all is silent. The two coffers – one holding Rena the other holding Harry, both in cryogenic hibernation – lie side-by-side in endless sleep.

> *Who made you glorious as the gates of heaven,*
> *Beneath the keen full moon?*
> *Who bad the Sun clothe you with rainbows?*
> *Who, with living flowers of loveliest hues,*
> *Spread garlands at your feet?*
> *Thou dread'st ambassador from earth to Heaven:*
> *Great hierarch tell thou the silent sky,*
> *Tell the stars and tell yon rising Sun, that Earth,*
> *With her thousand voices, forever praises God.*
> *Coleridge*

EPILOGUE

And so a super-hero is born, the GREY MAN, Harry Mandrake: titanic, invulnerable, colossus, and m*essenger for the Gods?* Pah! More like, *bookies runner for the Gods.* Fleet of foot, defiant, champion of excellence, style, and good taste; brave, heroic, fearless... Well, perhaps not 'fearless'... you know Harry.

So then... what will the modern-world hold for Harry and his *starchild*? We'll have to wait and see... see if Harry communicates with me from beyond the Badlands.

P J Searle

pjsearlebooks@gmail.com

www.ingramcontent.com/pod-product-compliance
Lightning Source LLC
Chambersburg PA
CBHW020629220526
45464CB00001B/73